Between the Tides

# Between the Tides

## In search of sea turtles

by George Hughes

First published by Jacana Media (Pty) Ltd 2012
Second and third impression 2016
Fourth impression 2023

10 Orange Street
Sunnyside
Auckland Park 2092
South Africa
+2711 628 3200
www.jacana.co.za

© George Hughes, 2012

All rights reserved.

ISBN 978-1-4314-0562-6

Cover design by publicide
Set in Sabon 10.5/15pt
Job No. 004035
Printed and bound by Inside Data, Cape Town

See a complete list of Jacana titles at
www.jacana.co.za

*This book is dedicated to Lee, Mitchell and Catherine
who shared much of the journey and sacrificed
a great deal to allow me to enjoy myself.*

# Contents

Maps ................................................................. 3
Preface ............................................................... 7

1 The Timeless Turtle ............................................ 11
2 Exploitation Until the Twentieth Century:
   A General Background ......................................... 17
3 The Eleventh Hour? ............................................ 23
4 From the Beginning ............................................ 30
5 Tools of the Trade – Part 1: The Humble Tag ................... 40
6 Tools of the Trade – Part 2: The Satellite Tag ................ 47
7 The Search for the Lost Decade
   – Part 1: The Nuclear Threat .................................. 55
8 The Search for the Lost Decade
   – Part 2: 'Isn't that one of your f---ing marked turtles?' .... 62
9 Amanzimbomvu to the Rescue and Other Tales .................... 71
10 'There's a *dog* nesting on the beach?' ...................... 80
11 'Sea turtles are not consummated in the market!' ............. 87
12 ORI and the Fairy Godfather .................................. 94
13 How Hatchlings Live and Die and Other Dangers ............... 101
14 Guiding Turtles to and fro and Other Wonders ................ 109
15 Aragonite Blue .............................................. 118
16 Politics and Turtles ........................................ 127
17 Conferences, Politics and Astonishment ...................... 135
18 Sustainable Use and the Great Divide ........................ 142

19 Enter the Mascarenes............................................150
20 Mozambique: The First Attempts.........................159
21 'A gross abuse of public funds!' ............................166
22 The Morning Chorus and Other Excitement ....................173
23 Of Rats and Mermaids ......................................181
24 Mozambique: The Far North ...............................190
25 Madagascar: Of Turtles and Tombs ......................198
26 Europa: A Superfluity of Animals .......................206
27 Tromelin Island ...............................................216
28 Among the Fairies: The St Brandon Islands ....................222
29 Beaches and Miracles .......................................230
30 The Sodwana Declaration .................................238

Photo Captions ...................................................246
Acknowledgements .............................................250

## The Mozambique expeditions

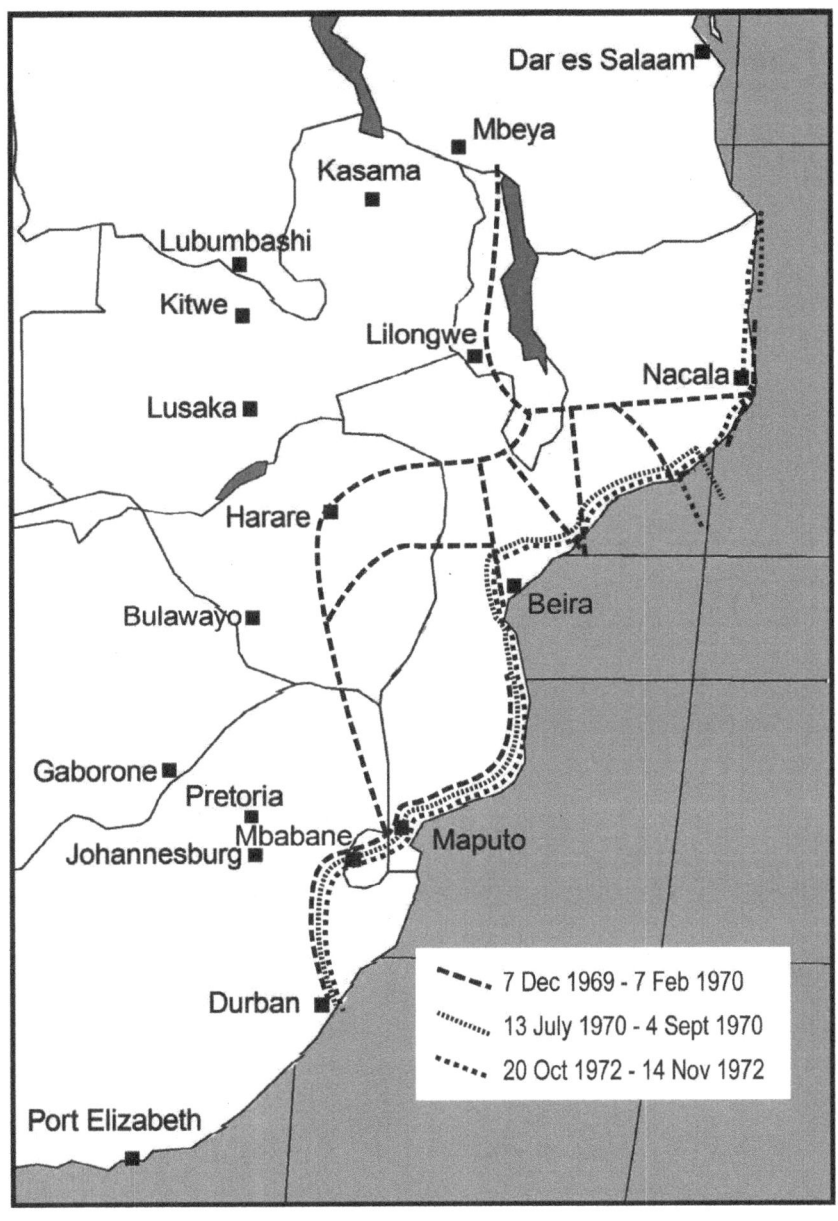

## The island expeditions 1969-1974

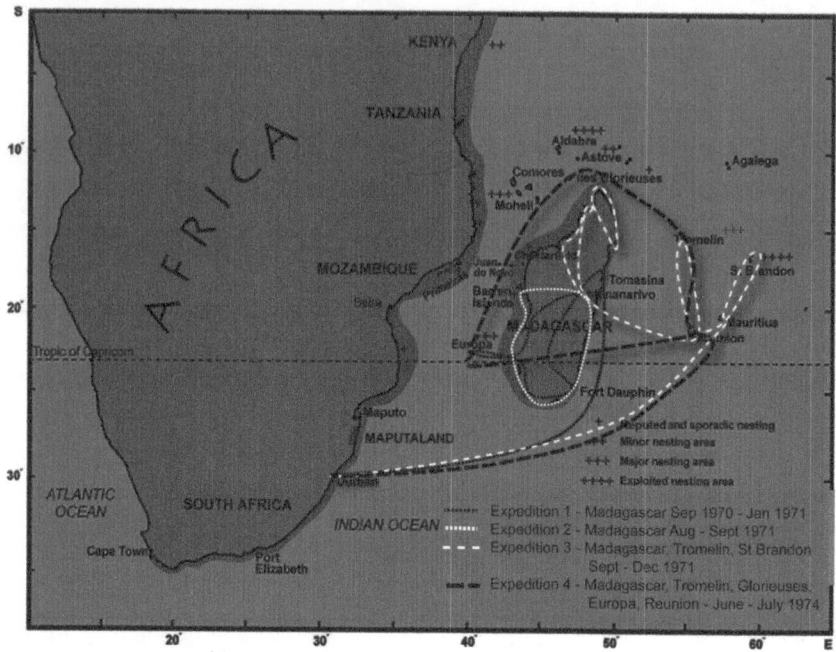

## Loggerhead migration routes

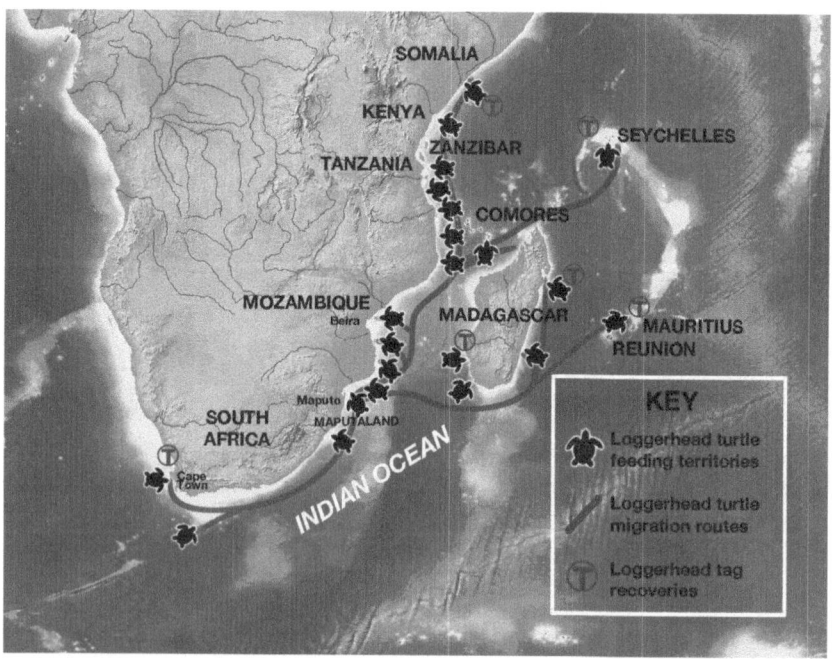

## Post-nesting movements of leatherbacks

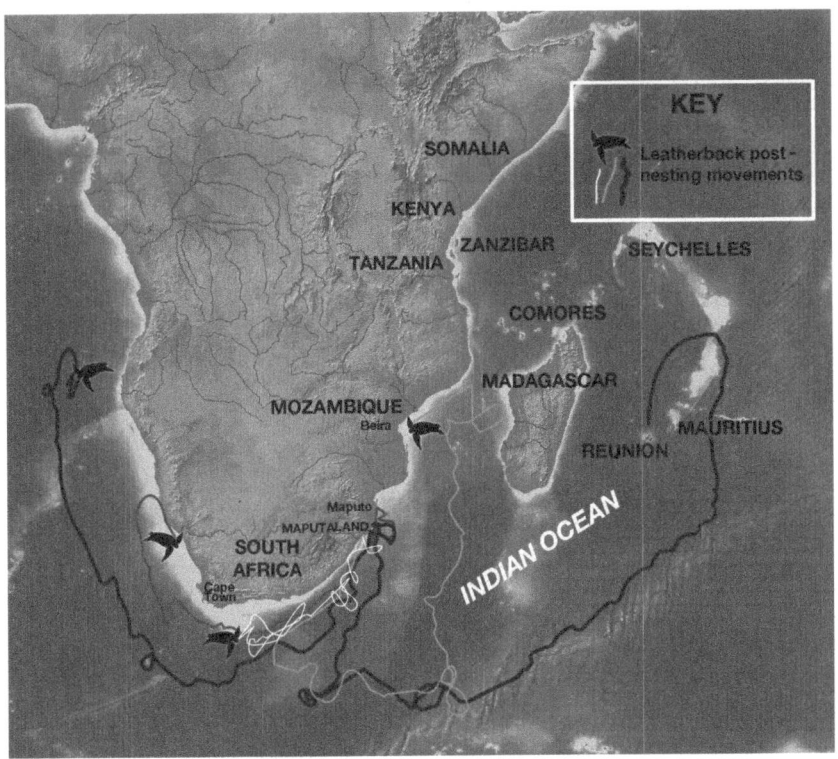

## The ocean gyres

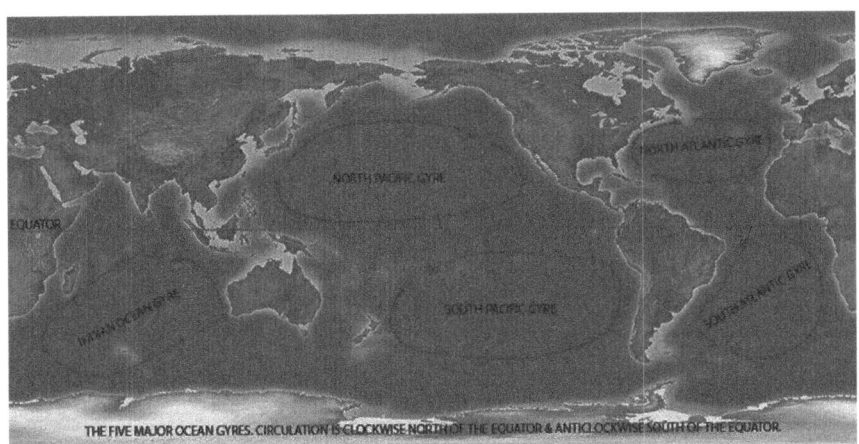

# Preface

Every now and again, upon meeting a person, you know instinctively that whatever that person intends doing will be done well. That proved to be the case when I met George Hughes in Durban towards the end of 1968. My first impression of him was of a fairly desperate young man who was nevertheless fiercely determined to carry out research in a field that he passionately believed was of fundamental importance – marine turtles. The obstacles in his way were formidable: no financial support, no base from which to work and no vehicle.

George had come to see me in my capacity as recently appointed director of the South African Association for Marine Biological Research (SAAMBR), a private non-profit company encompassing the Durban Aquarium and Oceanographic Research Institute (ORI), which functioned as a research facility of the University of Natal. In retrospect, it is questionable whether he came to the right person or institution at that particular time. SAAMBR was still reeling from the loss of its charismatic founder-director, Dr Davis Davies, who had died in a motor car accident. The original concept on which SAAMBR had been founded, namely that profits from the aquarium were to fund the ORI research programmes, had not proved feasible. The acquisition of outside funding for research had thus become necessary, but this too was fraught with difficulties. The ORI staff, many of whom hoped to use their research results for achieving master's or doctoral degrees through the University of Natal, were understandably very uncertain of their future.

My job was to get SAAMBR back onto an even keel and I was staggering under the huge load of responsibility. So, when George approached me, my first reaction was, 'As though I haven't got enough hay on my fork'. But when we got to chatting I discovered a man with a quick brain and remarkable insight. I told him about the crisis burdening our organisation and of my conviction that, if we were to pull through, we would have to expand our research from studies of sharks and exploitable fish resources to programmes that would enable us to offer scientifically based advice on wider aspects of the management of the coastal zone. George immediately responded that marine turtles are not just dependent on the sea, but are equally dependent on beaches for egg-laying and producing hatchlings. 'That being the case,' he asked, 'what better research programme do you require for providing advice to provincial authorities on land–sea interactions?'

I agreed, perhaps a little reluctantly, as I was acutely aware of the fact that coastal ecological processes and land–sea interactions go much further than meeting the reproductive requirements of marine turtles. Nevertheless, George had convinced me that the marine-turtle research he was proposing really had to be done. He and I both knew that I was not in a position to offer financial support, let alone a vehicle. What I was prepared to promise, however, was academic supervision and a place to work, which was initially a modest table and chair in a corner of the ORI library.

Bearing in mind my other problems, I also somewhat rashly undertook to seek funds for marine-turtle research under the auspices of the ORI and, if at all possible, to find a suitable vehicle. What I required urgently was a written motivation powerful enough to convince outside funding agencies of the vital importance of marine-turtle research, perhaps as indicators of the health of marine and coastal ecosystems and of their wider management requirements. George sat down at the library table and handed me a document a day later that was one of the most convincing I had ever read.

Suffice it to say that we were successful in finding at least some of the required funds, that he augmented these through other applications and that the Southern African Nature Foundation (now WWF-SA) provided funds for a second-hand Land Rover. He aptly named this vehicle 'The Panda Wagon'. George immediately started working with incredible energy – studying literature, doing fieldwork on remote beaches, travelling, writing and lecturing. Amazingly, he placed the manuscript of his PhD thesis, 'The Sea Turtles of South East Africa', on my table scarcely four years later. His research was of central importance as part of the overall activities of the ORI, which were thankfully productive again.

What I was not to know in 1969 was that George's unflagging dedication and energy would result in him leading marine-turtle research in South Africa for over 40 years, becoming a world-renowned authority in his field and playing a pivotal role in turtle conservation on a global scale. This is remarkable, as much wider and highly demanding responsibilities had in the meantime been placed on his shoulders, inter alia, as senior staff member and eventually CEO of the Natal Parks Board. In later years, especially when I was serving as CEO of WWF-SA, I did not hesitate to draw on his wide experience and knowledge – in a way, I made him pay for that Panda Wagon.

It is a privilege to have been asked to write the preface to this remarkable book and I have done so with great pleasure. The chapters that follow tell a story that should not be missed. It tells of determination and dedication, of hardship and joy, of frustration and the overcoming of many difficulties and problems. The reader will be taken to remote and romantic beaches and far-flung corners of the globe. The text is riveting and the story is told with the enthusiasm and humour typical of George. Above all, this book represents a valuable contribution to the national and international annals of marine and coastal science.

*Allan Heydorn*
*July 2012*

CHAPTER 1

# The Timeless Turtle

The privilege of walking along exotic beaches devoid of man on both moonlit and dark nights is given to very few. The excitement and wonder of finding, between the tides, sea turtles intent on nesting, is given to even fewer. I have been one of the lucky few.

One of the great myths about turtles and tortoises is that they attain great age, possess great wisdom and represent continuity, if not perpetuity. Many Eastern religions regard turtles as the foundation of the earth and such myths have been perpetuated by contemporary authors, such as Terry Pratchett with his Discworld series.

As far as sea turtles are concerned, there is proof neither of great age nor of great wisdom. Of the many thousands of turtles I have observed all over the world, I have been struck by the fact that very few display any signs of age. Size is certainly no indicator and, although there have been attempts to calculate age from rings in the humeral bones, it has proved impossible to ascribe turtles a scientifically defendable age from external characteristics. What are often visible are the scars from their time as hatchlings or as a result of later encounters with large marine predators such as sharks or killer whales – or sheer bad luck in trying to negotiate rocky approaches during periods of rough seas – but these are not any indication of age. Most researchers are confident, however, that painstaking research will reveal exactly how long sea turtles may live.

The late Prof. Archie Carr recorded some apocryphal tales of turtles found with inscriptions carved into their carapaces, which suggested that they had endured at least 85 years since being inscribed, but modern estimates fall somewhat short of that. At present we are certain, for example, that Maputaland loggerhead turtles take about 24 years to reach nesting maturity after entering the sea as hatchlings. We also know, from tracking females that return to Maputaland season after season, that these turtles can re-migrate many times – in individual cases these migrations occasionally span periods of some 25 years. Although there is nothing in the anatomy of nesting turtles to suggest age, our results would suggest that 50 years is not unreasonable for a female loggerhead and, considering the dangers facing a turtle every time it travels to nest, most turtles would do well to achieve it. But we may be well off the mark.

Many years ago, while walking along the Tongaland (which became known as Maputaland after 1975) beaches on a fairly bright evening, my fellow Thonga turtle-trotter, uMsombuluko, and I spotted a female loggerhead emerging from the sea. She was a black spot lying on the sparkling wash zone with just a hint of light reflecting off her still-wet carapace. Normally one would wait until the turtle was clear of the wash zone and then approach the animal from the rear to check for tags. She appeared to walk with patient deliberation and some awkwardness, however, so we decided to let her get a little higher up the beach before approaching. Perhaps we were just tired and grateful for a rest, as beach walking at neap tide was energy draining and really hard work on some nights.

We waited a long time, patiently watching her as she cleared the wash zone, entered the dry sand above the high-tide mark and eventually stopped well up the beach where her instincts told her that her eggs would be well clear of even the highest of high tides. She started to dig her body pit.

We laboured to our feet and quietly moved towards her from the rear, careful not to disturb her, and gently felt for a tag in order to ascertain whether she had been seen either this season or in any previous season. She bore neither a tag nor any trace of a callus suggesting that she had lost a tag. She was, as far as we were concerned, a new nester, perhaps coming ashore for the first time in her life. Untagged animals always caused a small frisson of excitement, so we eagerly dropped our packs and began to sort out the tagging and measuring gear preparatory to adding yet another unique female to our schedules of tagged animals. Having prepared our equipment, we switched on a torch and received a dramatic shock; she was the first and only loggerhead turtle I had seen, or have been destined to see, that displayed all the characteristics of great age – she was clearly a very, very old turtle.

She did not appear to have any spare flesh. Her skin hung in wrinkles of a size and depth that I had never seen before and at every exposed part of the body the skin drooped in long loose folds. What was more, the skin was free of the normal clusters of acorn barnacles so common on the majority of turtles. It was as if there was simply insufficient body to the skin for the barnacles to maintain a solid grip.

She ignored the pair of us and the torchlight, determinedly digging her body pit with steady sweeps of her fore-flippers until she reached the depth at which the top of her carapace lay level with the surface of the beach. We watched in fascination. This was no normal turtle and we did not want to disturb her, so we made no further moves to implement our normal procedures. Her every feature demanded respect and we somehow felt that this unique moment was to be savoured to the full.

Every tendon in her neck stood out sharply under the skin. Every head movement brought to the surface an array of tendons apparently unattached to visible muscle, but seemingly operated from within the carapace. It was like watching a marionette, but a marionette with almost no eyes. Close examination showed that her eyes were so deeply sunken into the sockets that they were barely visible and it was not possible to see the brown pupils so common to the species. It became clear that her every movement was laboured and each was taking a greater and greater effort.

At first sight, and clearly unkindly, we had thought her the most decrepit turtle we had ever seen. We thought that she was either very sick or suffering the after effects of serious wounds. Her probably extensive migration must have exhausted her energy reserves and she was clearly on her last legs. Our first impulse was to feel pity, for this animal was in a sad state.

How wrong could we have been? This was no sick animal worthy of pity but an ancient turtle whose lifetime became an object of excited speculation as we observed that her flippers, having moved steadily to and fro over goodness knows how many years, had worn half-moons in the outer edges of the carapace. At the deepest point, at least 10 centimetres of carapace had been worn away. We had never seen this before in Maputaland. The only explanation was that this turtle was of a great age, extensively travelled and making yet another visit to the nesting beaches to ensure the survival of her genes and her species. She had survived odds that are dazzling to contemplate. We sat in awe, amazed at her efforts, against greater and greater odds, to continue her line.

Speculation about where she had travelled, what dangers she had faced and how many times she had nested was both exciting and reverential; we watched her slowly complete her nesting hole and then settle, almost with an audible sigh, to lay her eggs with her now-exhausted rear limbs straddling the hole. After an unusually long time, she started the internal contractions that brought the eggs down into the hole in bursts of two to four. Long pauses followed each expulsion of eggs and it took her twice as long as normal to finish laying her clutch – which comprised more than 100 table-tennis-ball-sized, soft white eggs – and then she rested.

By now we were emotionally tied to the old female and we waited without impatience for her to fill the hole and disguise the nest site. We were enjoying the moment, fully realising that we were unlikely ever to see a similar animal again, when we were both surprised and shocked. Instead of filling the hole with sand, a series of spasmodic convulsions shook her and after some minutes we were amazed to see pink flesh emerging from her cloaca. By now we were riveted to her every move and minutes stretched into tens of minutes while she evicted her entire oviduct into the egg hole. Then she stopped and we sat there bewildered, having never seen or heard of such an occurrence before. We waited to see what would happen next.

After a lengthy pause, she reverted to the standard loggerhead practice of carefully lifting sand with her hind-flippers and placing it gently into the nest hole, sliding from side to side as she did so. After about ten such moves, she once again stopped, as if contemplating the next step of the process, before pressing down firmly on the sand to compact it and provide added protection for the eggs. Oblivious of the gradual disappearance of her oviduct, she carried on with her normal nesting process. With weary fore-flippers, she made a weak attempt to disguise the entire nest site. By now, however, she was obviously exhausted and after a long pause she clearly decided that she could do no more. She pulled herself toward the sea, tearing her oviduct from her insides and effectively ending what had obviously been a long lifetime of dedicated service to her species. It was a dramatic moment as we stood watching her walk back to the sea, probably for the very last time. There would never be another instruction from her hormones either to mate or to migrate; no instinctive urge would ever again cause her to leave the relative safety of her feeding grounds to emerge onto a beach.

It was a sobering experience. We wordlessly decided against tagging her, or even measuring her. She did not deserve any indignity. We slowly followed her down to the sea. We stood in the water and we watched as the first small waves swept the beach sand from her head and carapace. Then she paused, as if to take stock of her surroundings, lifted her head and peered at the sea horizon, slowly crawled forward and was swept from view by a large wave. She was gone.

After a contemplative silence, we went back to the nest and carefully removed the oviduct from the nest site, ensuring that no part of it remained with the eggs. Taking perhaps more care than we normally did, we covered the nest and disguised it thoroughly. Never had a clutch of eggs better deserved a successful hatching. I am sure that many of the hatchlings that eventually emerged are already continuing the duties of their mother.

In all the years that have passed since that night, I have never again witnessed such an incident and neither have any of my turtle colleagues in many parts of the world. I reflect on what I saw with some humility. The life of a productive nesting female sea turtle is not an easy one and although I have watched the emergence of countless thousands of hatchlings, I have never felt so close to a sea turtle as I did to that loggerhead female on that quiet night in Maputaland. She had completed her nesting ritual, not just for that clutch of eggs, not just for that night and not just for that season, but for ever, and departed to seek out the rest of her life freed from the vicissitudes of breeding. My own judgement is that she had done an extraordinary job and deserved some years of peace for having made her unique and full contribution to the continuation of her species.

It would have been nice to know exactly how old she was, but that will forever remain a mystery. Since then, however, many of us have laboured long and hard to describe the life history of sea turtles – and this book describes our endeavours.

CHAPTER 2

# Exploitation Until the Twentieth Century: A General Background

Sea turtles are essentially prey animals and one of their basic survival strategies has been to produce large numbers of eggs per female for as many seasons as the female can manage. This is no trivial strategy, as most females produce over 100 eggs per clutch and some, like leatherbacks, can lay ten clutches per season. This is a total of over 1,000 eggs in a single four-month period. This immense labour can cause a female to lose almost 25 per cent of her body weight and drains the animal's resources to a point where many appear unable to endure more than one season of nesting. The strategy has, however, been very successful and, although it was developed well before the coming of humankind, it has allowed all seven extant species to survive one of the most savage attacks by a single species on another ever recorded.

In many parts of the world, early man discovered that the annual gatherings of nesting sea turtles were simple and threat-free opportunities to obtain high-quality protein. They killed the odd nesting female and collected and consumed the easily available eggs. While these early hominids were serious predators, their numbers were small, their lifestyle was nomadic and the impression that they made on sea-turtle populations was negligible.

They were not the only predators, however, and turtle rookeries attracted the attentions of many other terrestrial predators, from jaguars in South America to racoons in North America and dingos in

Australia. In South Africa, side-striped and black-backed jackals visit the beaches nightly to dig up eggs and are joined by the Nile monitor and, more recently, the ratel or honey badger. All such predators were comfortably catered for by the turtles' strategy, which has been designed over millennia: overwhelm your enemies by sheer numbers.

When western man started to 'discover' new lands and conquer the oceans, it was necessary to overcome difficulties and threats hitherto never experienced. Sailors, right up to the end of the eighteenth century, suffered from the effects of a diet lacking in vitamins and fresh food. The sea turtle entered into this scenario and the previously tolerable influence of early man became life threatening and eventually wiped out some of the largest turtle rookeries in the world. Exploitation gradually grew to industrial proportions, which the turtles' survival strategy had never been designed to combat.

Sea-turtle rookeries were identified by the early explorers and their locations were noted as destinations for replenishing food stores. To make matters worse, they learned that turtles could last for months if given a modicum of care – such as being kept on their backs in the shade or in the hold and watered down regularly – and could be kept to be eaten on demand. The huge numbers of turtles at some nesting sites inspired understandable awe, so much so that Christopher Columbus named one Caribbean island group Las Tortugas (the Turtle Islands), which is now known as the Cayman Islands. The Cayman Islands eventually became famous as a stopover for ships plying the waters of the New World. The good news was that the turtles cost nothing. The bad news was that the system could not endure and there was worse to come.

Slavery in East and southern Africa was the domain of Arabs throughout the eighteenth century, and indeed well into the nineteenth century, until the British blockaded the main export port of Zanzibar in 1873. In West Africa, however, the slave trade became industrial and every Western nation with a respectable trading fleet indulged in slavery, which led to the development of what became known as the 'Slave Triangle'. European ships bearing materials manufactured in Europe would sail to West Africa where they sold their cargoes for

## Exploitation Until the Twentieth Century: A General Background

cash to buy slaves or traded the materials directly for them. These evil transactions grew in proportion with the prosperity of the New World until millions of Africans were being exported through the slave markets of West Africa such as Gorée Island in Senegal, Ghana, Nigeria and Sierra Leone. Following the successful sale of the surviving slaves in the New World, the ships would load up with products such as sugar, tobacco and cotton and return to Europe, thus closing the Slave Triangle.

Feeding a few sailors on sea turtles gathered on rookeries was one thing, but providing free food for slaves packed into ships in inhumane numbers was another. Slaver after slaver called at Ascension Island in the mid-Atlantic to collect turtles to feed their captives. The negative impact of the slave trade soon began to manifest itself and the numbers of green turtles nesting on the island declined sharply. By the time slavery was outlawed, the island was no longer regarded as a source of turtles.

The colonisation of many remote islands, such as France's Reunion Island (originally named Ile de Bourbon), Mauritius (Ile de France), Rodriguez, the Seychelles and many others, led to increased shipping and the increased exploitation of nesting sea turtles. The growing human populations on these islands, who regarded nesting turtles as free food, exacerbated the problem until these large colonies of sea turtles also disappeared.

Accompanying the newly resident colonists on both mainland and island were domesticated dogs and pigs. Desertions by dogs led to enormous damage to nesting colonies of turtles as hunting packs of feral dogs wandered the rookeries at night, digging up eggs and consuming hatchlings en route to the sea. Feral pigs, escapees from farming ventures, found turtle eggs similarly tasty and an easily accessible menu item.

As seagoing trade expanded and became more efficient, ordinary trading voyages, as well as slavers closing the Triangle, would occasionally convey the odd surviving turtle back to Europe. Any such survivor was sold on the open market for very high prices. The prices, mainly for green turtles, were so exorbitant that they would only attract the attention of the very rich. Thus began the tradition that sea-turtle products regularly graced the tables of the wealthy classes and those of noble birth as meat or, more famously, soup. As the domestic consumption of green-turtle products grew in popularity, so did the pressure on wild populations. Trade in turtles became a mainstream activity and turtles took their place as a commodity imported into the Western world.

The Lord Mayor's annual banquet in London has served green-turtle soup for generations. (Strangely enough there was an exception to this annual event in 1970 after a visit by three enthusiastic young biologists, including me, who appealed to the better nature of the Lord Mayor's office and carefully explained the problems facing sea-turtle populations. The exception lasted one year.) Creamed green-turtle soup, made by Lusty's of London, has been obtainable from Fortnum & Mason for generations, and throughout the Second World War was habitually enjoyed by Sir Winston Churchill. This enthusiastic passion for turtle products was shared by the affluent in France, West Germany and, of course, the United States, where the Americans developed their own unique styles of preparing and eating turtles. They even manufactured and sold 'turtle burgers'.

As with all commodities that are the envy of the less wealthy, the intensive hunting of green turtles did not cease with the end of slavery. The objectives of the hunts did, however, change from feeding sailors and slaves to commercial trade and, as a result, many huge nesting assemblages were accosted and brought to total ruin.

Ascension had already been ravished, but in the New World commercial hunting soon eliminated all green turtles nesting on the beaches of the Cayman Islands. In order to replace this resource, a fleet of specially designed sailing vessels called turtle schooners was built on Grand Cayman. For over a century, they collected turtles along the Atlantic shores of Central America where they assisted in bringing other nesting populations to the brink of extinction. The last of the schooners was broken up only in the 1970s.

In other parts of the world, exploited turtle nesting sites were plundered without any thought given to the survival of the stock.

## Exploitation Until the Twentieth Century: A General Background

The early companies simply could not grasp that the resource could be diminished. They saw, in some cases, thousands of animals nesting every night and never considered the fact that their thoughtless slaughter would lead to permanent damage to, if not the extinction of, their endeavours. But that is what happened. Permanent processing settlements on the large turtle rookeries of Aldabra and the Seychelles, Europa Island in the Mozambique Channel and many others, including the turtle canneries in Somalia and on Heron Island, Australia, eventually failed as the numbers of animals coming to nest dwindled.

This cavalier and destructive approach was not restricted to green turtles, although they are the most commonly eaten. They are herbivores and their meat is excellent. Soup is made from the dried cartilage of the carapace (upper shell) and plastron (lower shell) edges, which are rendered into a thick creamy solution. This soup goes beyond the unique; it is without doubt one of the great tastes of the world and is highly nutritious. Green-turtle oil is invaluable and was once a key ingredient in women's toiletries, creams and balms.

The hawksbill turtle, regularly seen in southern African waters, is the source of the most beautiful tortoiseshell in the world. It has been highly valued as a decorative medium as far back as, and probably well before, the reign of Cleopatra and is used in the manufacture of personal items such as jewellery, mirrors, combs and spectacle frames. Tortoiseshell has also been widely used for inlays in expensive furniture items. The demand for tortoiseshell expanded dramatically in the nineteenth century and throughout the tropical world intense pressure was brought to bear on this species. Madagascar became a prime source of exploitation and tens of thousands of kilograms of tortoiseshell were exported each year, until the collapse of the industry at the end of the First World War.

The demand for this beautiful product has not disappeared altogether, however. The Japanese have a centuries-old cultural relationship with the working of tortoiseshell, known as *bekko*, which has developed into a singular and truly stunning art form. Even today, an annual gathering is held in Nagasaki where artisans compete for recognition in the artistic and innovative use of tortoiseshell. Numerous establishments that work solely with tortoiseshell are still operating in Japan and their products are still highly valued and sought after. Need I add that the products are expensive, and becoming more so, because there is no legal international trade in hawksbill-turtle products. The hawksbill, as well as all other species of sea turtle, is listed on Appendix I of the Convention on International Trade in Endangered Species of Wild Fauna and Flora (CITES).

The dramatic reduction in the numbers of green and hawksbill turtles around the world, and the near disappearance of the olive Ridley turtle in Mexico in the twentieth century as a result of the demand for turtle leather, are all prime examples of the 'Tragedy of the Commons'. No one took responsibility for the turtle rookeries and no one paid any attention to mismanagement and ruthless exploitation. It was good old-fashioned economic theory: benefit from the resource while it is there.

Humankind did not learn from these early examples of irrational and irresponsible extraction of resources, and recent examples of exploitation, without sound science as a basis for the endeavour, can be seen in negative effects on many species other than turtles. The most recent noteworthy disasters have been the cod fishery on the Newfoundland Banks and the threatened extinction of the tuna stocks. It has been said that no major fishery in the world has been efficiently managed. One of the worst examples of mismanagement has been the whaling industry, with which South Africa was associated until the 1970s when the whaling base at the Point in Durban was closed down.

The presence of sea turtles in southern African offshore waters had long been noted, and sea turtles were, like the land mammals, eaten occasionally when available. But, by the end of the nineteenth century, sea turtles had dropped from public view and their status and future was ignored. No nesting grounds had been identified along the South African coastline. So, as far as South Africa was concerned, sea turtles did not exist.

CHAPTER 3

# The Eleventh Hour?

By the end of the nineteenth century, there was a growing concern in South Africa, even among political figures, that the once super-abundant populations of large mammals and birds, which had made such an immense contribution to the economy of the region – and to the survival of so many of its peoples, both indigenous and immigrant – were on the point of disappearing completely. For over 250 years, the large-mammal populations in South Africa had been exploited with no thought given to the possible exhaustion of the resource – the same scenario that led to the destruction of many sea-turtle colonies across the globe.

All credit must be given to the Natal colonial government for taking the bold step of setting aside large tracts of land in Zululand (Hluhluwe, Umfolosi and St Lucia game reserves were proclaimed in 1895) and the Drakensberg mountains to protect their wildlife. These protected areas, set aside primarily for large mammals, are the longest surviving in Africa. The example was rapidly emulated elsewhere (Sabi Game Reserve was declared in 1898 and renamed the Kruger National Park in 1926) and in Natal relatively generous funds were made available to care for and expand protected areas. These moves attracted a great deal of opposition, but legislators held firm and the protected areas across the globe gradually achieved great respectability and received public support. The age of conservation had begun. Not for sea turtles, however.

As far as legislation was concerned, it appeared that sea turtles in southern Africa were extinct at that time. Well, perhaps not quite. In South Africa, where turtles were regarded as little more than an occasional oddity, provincial conservation officers noted that there were fewer turtles being encountered. In 1916 the Natal provincial government proclaimed modest legislation, which provided some protection to the turtles encountered on the Natal coast. It attracted little interest other than the fact that the legislation was associated with measures forbidding the shooting of dolphins near Durban, which was a far more visible practice and one that many citizens of Durban had found offensive. Turtles were not to receive greater legislative protection for another 31 years, when Provincial Act 34 of 1947, which brought the Natal Parks Board into existence, was passed.

Elsewhere in the region, the situation was little different. The ruthless exploitation of hawksbill turtles for tortoiseshell continued unabated in Madagascan waters in the early twentieth century and numbers were dwindling rapidly. Green turtles were being exploited excessively on Aldabra Island and traditional killing continued apace on the granitic islands of the Seychelles. A small settlement had been established on Europa Island and the slaughter of the island's turtles swiftly brought their numbers to a point where further exploitation was senseless. Not only were the settlers unable to harvest enough turtles, but the harsh, waterless island soon claimed the lives of many of the settlers as their sisal crops, which were chosen to supplement the dwindling turtle harvest, failed to meet expectations.

There were, however, stirrings of conscience and a gradual realisation that sea turtles could not be exploited indefinitely. This prompted Madagascar (then a French colony) to enact laws in 1923 to end the exploitation of turtles on the outer islands, such as Europa, and they restricted what could be hunted or harvested in mainland coastal waters. It was as late as the 1960s, however, that the Seychellois took steps to protect their fast-dwindling sea-turtle resource, as did others such as the Comoros, Mozambique, Mauritius and Reunion Island. Most of the legislation was never enforced with any serious conviction, however, and turtles continued to survive in the region almost by default. There was a near total official disinterest in sea turtles throughout the south-western Indian Ocean. Only Europa Island benefited immediately from the legislation in 1923, as all settlers were withdrawn from the island. This created an opportunity for the nesting population to recover and eventually become, once again, the largest green-turtle rookery in the region. It is not known how many green turtles were killed during the exploitative period, but the mounds of turtle bones still present and reported on in the early 1950s bear testimony to the immense slaughter.

On the rest of Les Îles Éparses – which include the islands of Tromelin, Juan de Nova and the Glorioso Islands, controlled from Reunion Island – and Reunion Island itself, no specific attention was accorded to sea turtles. The beaches of Reunion had long since been stripped of nesting turtles and the islands were too distant to bother about. Only after the Second World War, and the establishment of meteorological stations on the scattered islands, was there the means to access these islands. Even then few visitors took the risk.

No practical steps were taken in Mozambique to curb the killing of turtles until the 1970s when the Paradise Islands were declared a protected area. This gave sanctuary to the northernmost nesting females of the loggerhead turtle meta-population, whose nesting grounds stretch from the Paradise Island National Park to the St Lucia Estuary in KwaZulu-Natal and along the southern coast of Madagascar. Outside the Paradise Islands protected area, however, incidental killing of nesting and migrating turtles in Mozambique continued virtually unabated until the beginning of the twenty-first century. Domestic killing of nesting or incidentally caught sea turtles continues along the Mozambique coast, even today, despite the excellent efforts of a small army of volunteer turtle-watchers along the 3,000-kilometre coastline.

At the northern end of the Mozambique Channel lie the Comoros Islands. Little was known about these islands, which were under French control, aside from the fact that coelacanths could be found there. In 1948 a coelacanth was eventually collected for scientific study, thanks to the energy of Dr JLB Smith and the cooperation of Prime Minister Dr DF Malan who made available an air force Dakota. Nothing appeared to be known about sea turtles there, however.

Along the south coast of Madagascar, the killing of loggerhead turtles has never ceased, although there are dedicated turtle conservationists active in the area today. The general disinterest of the official authorities has not improved and prosecutions for killing turtles remain a rarity. This unfortunate state of affairs is not entirely due to inefficiency or official disinterest. Many coastal inhabitants survive off the sea, with no other source of sustenance or income, and there is widespread official reluctance to deprive people of their already meagre means of survival.

In the 1960s the Seychelles started to take firm steps to control the killing of turtles in the archipelago. This almost led to riots among the general population and the legislation was withdrawn after only a few years. The indigenous Seychellois were not happy with the laws, which threatened what they regarded as their traditional rights to harvest green turtles for food and hawksbill turtles for tortoiseshell. Artisanal tortoiseshell products were openly sold in the streets of Mahé, the capital of the Seychelles, as recently as 1997. Shortly thereafter, however, all traffic in sea-turtle products was forbidden and strenuous efforts were made to encourage the long-term survival of sea turtles. The private sector has played a positive role in the granitic islands. On Cousine Island, Mike Keeley, the South African owner since 1992, has not only strenuously enforced protection laws, but has employed biologists, two ex-colleagues and good friends Dr Orty Bourquin and Peter Hitchins (both ex-members of the Natal Parks Board), to monitor nesting turtles. Their research has revealed dramatic growth in nesting numbers. The driving force for turtle conservation in the Seychelles as a whole, however, is Dr Jeanne Mortimer who has dedicated much of her life to improving the future of turtles in this important island group.

A similar story of official disinterest unfolded in Mauritius. All of the turtles nesting on the main island had disappeared early in the nineteenth century and, despite the legislation passed to protect them, sea turtles on the outer islands also fared badly. As late as the 1970s, on the St Brandon islands (the Cargados Carajos Shoals) some 400 kilometres north of Mauritius, contract fishermen were employed by the Mauritius Fishing Development Company to work a dried-fish concession. They enjoyed free access to the turtles nesting on the islands and the company itself brought regular cargos of green turtles back to Port Louis, where they were openly sold to an appreciative market. By the end of the twentieth century, however, all such killing had been stopped and some of the islands in the group have been set aside as protected areas.

Serious wildlife conservation really started after the Second World War, as the United Nations began to gain confidence, and this led to the establishment of a robust series of conservation-based global institutions. These included the International Union for Conservation of Nature (IUCN); the United Nations Educational, Scientific and Cultural Organisation (UNESCO), which embraced the concept of World Heritage Sites (both natural and cultural); the Ramsar Convention on Wetlands (Ramsar is a massive wetland in Iraq where the first gathering took place to draft a treaty for the protection of wetlands); the Convention on Migratory Species (CMS); the Declaration of the Convention of Biodiversity; and literally hundreds of NGOs, the largest of which has been the World Wide Fund for Nature (WWF). Last, but not least, the Convention on International Trade in Endangered Species of Wild Fauna and Flora (CITES) came into force in 1973.

Under the umbrella of these large global conventions, a series of more focused bodies of conservationists developed. The IUCN's Marine Turtle Specialist Group, which was formed to get recognition and protection for the world's populations of sea turtles, more or less led the pack. The force behind the establishment of this group was Prof. Archie Carr from the University of Florida in Gainesville. Archie's drive, enthusiasm and leadership stimulated a worldwide movement of which he would be enormously proud if he were alive today. He would be proud, but still concerned.

The IUCN managed to amass only 12 scientists who were actively working on sea-turtle research across the world and invited them to attend the first meeting in Morges, Switzerland, in March 1969. It was my rare privilege to be one of them, representing the Natal Parks Board and South Africa. How I got there happily connects with a flow of serendipity that could not have been more propitious had I planned it deliberately.

More than four of my post-teen years were spent as a Natal Parks Board game ranger at Giants Castle Game Reserve in the Drakensberg mountains. During those halcyon days, I had started to take more than a passing interest in scientific method and data collection on numerous animal groups, such as small mammals, snakes and frogs. Towards the end of my stay in the Drakensberg, this expanded to the recording and interpretation of San rock art. These endeavours were exciting and rewarding and I was eventually persuaded to pursue formal studies at university by a young lady friend. As I had no money, I had to work for another 18 months before garnering enough to pay for part of my studies. In addition, my small mammal research had earned me a modest reputation and some excellent friends, one of whom was the late Dr John Pringle, then director of the Natal Museum and himself a small mammal expert of note. Dr Pringle was a wonderful mentor who provided me with a tremendous amount of guidance and managed to secure for me the only Wildlife Society scholarship available (the only previous Wildlife Society scholarship had been awarded to a fellow Natal Parks Board game ranger, Ken Tinley, now Dr Ken Tinley, resident in Western Australia). To put things into perspective, this generous grant was made up of R150 for the first year of my studies. If I passed I would receive an additional R150 for the second year and, again, if I passed, a final instalment of R200. This made a grand total of R500 payable over three years.

Although I was ignorant of it at the time, having just decided on academic study, a report from the National Department of Bantu Administration and Development was received at the head office of the Natal Parks Board. It stated that sea turtles had been found dead on the beaches of Maputaland, south of the Kosi estuary mouth and the Mozambique border. It was suggested that these turtles had been killed while nesting. The Maputaland region of Natal had been virtually unexplored at this stage, aside from one Wildlife Society expedition to the area in 1947, whose report did not mention sea turtles.

This news came as a mild shock to Mr Peter Potter, assistant director of the Natal Parks Board, who was a keen coastal angler and had never heard of turtles nesting in South Africa. The thought excited him enormously and he immediately arranged for a staff member, Ranger Hennie van Schoor, to set up a protective presence on the Maputaland beaches near Bhanga Nek, not far from Lake Hlange in the Kosi estuary chain. He employed two students from the University of Natal, Humphrey McAllister and John Bass, to join Hennie and to try to establish how serious the poaching problem was, how many turtles were visiting the beaches and when they were making their visits.

Peter, Hennie, Humph and John started what became known as the 'Turtle Survey', which launched sea-turtle conservation as a serious endeavour. Viewed from 2012, the survey was to have enormous benefits for sea turtles in the region. The first survey was carried out between 1 December 1963 and 31 January 1964. It was during that period that I decided to go to university and two seasons later, as a vacation student, I joined the survey, never suspecting that turtles would dictate a large part of my life for the next 48 years.

*Bhanga Nek Cottage, the heart of the Turtle Survey, as it appeared in 1973. Sketch by the author.*

CHAPTER 4

# From the Beginning

In 1963 Hennie, John and Humph started from a position of almost complete ignorance as far as sea turtles in South Africa were concerned. They had no idea what they were looking for and were astounded, on their first nights on the beaches, to find not one species of sea turtle, but two. Having only ever heard of green turtles (one of them claims), they were somewhat confused and thought that loggerheads might be greens. Their knowledge of leatherbacks was unfortunately limited to having seen a picture in a geographic magazine, but after a couple of weeks, and some enquiries, they settled on loggerheads and leatherbacks.

When they visited the Oceanographic Research Institute (ORI) in Durban, (the research institute of the South African Association for Marine Biological Research [SAAMBR], better known as the Durban Aquarium), they confirmed that South Africa had two species of nesting sea turtles. These were the modest red-brown loggerhead turtle, *Caretta caretta*, the carapace of which can exceed a metre in length, and the leatherback turtle, *Dermochelys coriacea*. The latter is the world's largest sea turtle, with seven longitudinal ridges along the black carapace, which is patterned with blue-white spots and rosettes, and can reach a length of nearly 3 metres. During the first season, they encountered 142 loggerheads and 16 leatherbacks. In comparison with the few beaches around the world on which data had been published, these numbers were very low and it seemed that both species were en route to extinction in South Africa. Hennie, Humph and John made the first call to have the nesting beaches declared a protected area, but this was some years in coming.

With their enthusiasm increasing daily, the first season was spent trying to ascertain what was going on and to end the illegal killing of the turtles. The general success that this first team achieved was measured in the negligible killing of turtles on the beaches while they were there and the beginning of the programme of tagging adult turtles that continues today (see Chapter 5).

To say that conditions on the beach in Maputaland were primitive is an understatement. Hennie, as the resident officer-in-charge of the Kosi Bay Nature Reserve, employed a group of local Thonga people to build a modest cottage on the beach in more or less traditional Thonga style, which served the turtle survey for many years. Apart from the odd crude extension, the thatched-roofed and thatch-walled Bhanga Nek cottage remained virtually unchanged, until the South African Air Force, in a friendly frame of mind, put a Super Frelon helicopter down on the adjacent beach and blew the roof off.

The cottage nearly didn't survive its second year. Despite being informed about the killing of turtles by the Department of Bantu Administration, there was a degree of hostility between the Natal Parks Board and most government bodies. When it was reported that an illegal structure had been built on the beach, the Parks Board was immediately instructed (rather spitefully in Hennie's view) to remove it and it was decreed that tents alone would be permitted. Hennie was a resourceful man, however, and, as if by magic, Corrie Nel, the commissioner-general of the Zulus, arrived at Bhanga Nek to see what was going on. The ever-charming Hennie invited him in for a few drinks and a good old-fashioned Afrikaner chat ensued, during which Hennie discussed the turtles and the Natal Parks Board's work. He ended dramatically with the news that, tragically, the cottage would have to be demolished the following morning. 'No,' said the commissioner-general, 'that will not be necessary. I will see that the problem goes away.' And it did. Hennie had saved the cottage.

For the first five years of the survey, the cottage served as accommodation for the resident Natal Parks Board officer and his family. For the first year, the cottage had no mod cons at all – no electricity, no running water, no bath, no shower and a wood-burning stove that probably started global warming. There was a gas fridge and a gas freezer, but, for those of you who have never lived with these demonic devices, be assured that they were both a blessing and a curse. When they worked well they were great, but when the wick failed or started to misbehave, the air filled with paraffin soot and a foul smell that became detestable. In the second year the ever-resourceful Hennie laid on running water from a tank high up on the dune behind the cottage, which had to be filled by hand with water from Lake Hlange. A long pipe terminated in a tap and shower head at the cottage. Alas, this fell into disuse almost immediately as the pipe was forever blocked by the body of some creature that had inconsiderately died in there.

Fortunately Hennie was not entirely alone. In those days, the Natal Parks Board had its own vehicle workshops in Hluhluwe Game Reserve and, in times of vehicle trouble on the beach, help was seldom called for in vain. Humph had a slight accident with the Land Rover which called for skills that even Hennie did not possess and it was the first of the many times that the mechanical staff came to the rescue. The Parks Board's mechanics, exemplified by Jimmy Pattenden, a truly wild mechanic famous for his skills and language, took real pride in getting to Bhanga Nek as fast as possible. He would sometimes work throughout the day, carrying out any necessary repairs to ensure that the patrols continued according to plan. On one occasion I was helping Jimmy remove, clean and replace an engine on a beach buggy (sand was always a problem for that particular model). It was a very hot and humid day and the main engine parts would not align properly. I had always regarded Jimmy as a taciturn and fairly non-communicative man, but, when a third attempt at realignment failed, the floodgates opened and Jimmy directed a string of curses at the inanimate engine parts that exceeded everything in my experience and which, I am sure, remains an unbroken record at Bhanga Nek.

The students had a hard time in those days too. For the princely remuneration of R2 per day, they had to live in army bell-tents erected on the beach (these were replaced by the more airy, but still hot, square tents in 1966). As virtually all sea turtles nest at night, the students had to work then and had to live in the close confines of the tent, or be exposed to the Thonga sun, during the day. Following the demise of the shower, if one wanted a decent wash, one walked a kilometre or more to Lake Hlange, where the water was fresh enough to work up a modest lather with soap. The walk was generally a bit of a nuisance and the fact that there were crocodiles and hippos in the lake was a bit of a deterrent. So, a good rub down in the easily accessible surf served quite well for the most part, and salt and sand became permanent features of one's daily life.

The crowning glory of the Bhanga Nek accommodation was the outdoor toilet. Situated higher up the slope in the forest behind the cottage, it had a large portal through which one, while sitting comfortably, could absorb the most beautiful view of the sea and northern beaches. On a calm day one could, dreamingly, see forever. Apart from some geographical shifts, almost imperceptible to the regular visitor, the toilet remains in place and with the same view to this day.

Water was very scarce at the cottage in the early days and there were only two ways to get it to Bhanga Nek from inland. The quickest was across Lake Hlange by boat from the Kosi Bay Nature Reserve. It was possible to bring a 44-gallon drum of very good quality water across the lake in the 5-metre aluminium boat and then by Land Rover to the cottage. In November 1966 when this was attempted, however, the staff members were surprised in the middle of the 6-kilometre-wide lake by a storm front that capsized the boat. Two outstanding and loyal game guards, Mbika and Onions Gumede, lost their lives and the officer-in-charge spent many hours in the water before washing ashore. After that tragedy, large drums of water were never transported across the lake again.

The other route to the cottage was a sand track from Maputa, the nearest village, some 20 kilometres away. In those days, Maputa had an excellent and normally well-stocked store, belonging to Ndumu

Limited, a police station, a hospital and nearby was a Catholic Mission called the 'Star of the Sea' and that was about it. (For some reason, unbeknown to me, in the 1980s the powers that be changed the name of Maputa to Kwangwanase and I shall henceforth refer to it by its modern name.) En route to Bhanga Nek, the sand track crossed the Malangeni River through a magnificent swamp forest whose gentle filtering kept the river water clear, clean and drinkable, but with the colour of Earl Grey tea. Some 16 kilometres from the beach, it was from this river that we would, as few times as possible, collect drums of water for cooking and washing purposes. This was seriously hard work, as the track was not an easy drive. The entire road was narrow, rutted, very sandy and uneven. Whenever one had to drive up a slope (actually a grass-covered sand dune), it required real driving skills to make it to the top. In those days there were no synchromesh gears in Land Rovers, so it was double de-clutching or one got stuck. Getting there with the empty drums was simply uncomfortable; coming back with the drums after filling them by hand was uncomfortable and unpleasant, if not dangerous. There were often staff on the back of the vehicle and there was the constant danger of a rope slipping and a loose drum injuring someone.

Fresh food and recently baked bread were always in demand and even when we could get to Kwangwanase there was often no fresh food to be bought. The 4x4 Magirus truck of Ndumu Limited could often not get through because of the bad roads. They were made inaccessible by the severe summer rains, which regularly caused the Pongola floodplain to disappear under water for weeks.

The second season, in 1964/5, saw the arrival of Orty Bourquin, a new student who joined Humph and Hennie, and they continued tagging. The number of turtles tagged was considerably more than the first season, with 223 loggerhead females and 13 leatherback females tagged. Orty certainly made his presence felt as he was a rabid naturalist who collected everything that moved and a great deal of material that didn't.

The 1965/6 season saw an increase in student numbers and a change of officer. Brian Stevens was installed as officer-in-charge and, with his wife Christine and young son, took up residence at Bhanga Nek. Mike Mentis and I took up the student role and our season started in early December. We were later joined by John Bass and Orty Bourquin, who had signed up for a second year.

John's arrival was dramatic. We lost a tagging tool and had sent urgent messages to the Natal Parks Board head office to send us a new one, as we were down to one. In response, John was despatched to join us earlier than expected and, because of the flooding of the Malangeni, he couldn't reach us by road and decided to wade across the lower end of Lake Hlange. This was not something he had done before and halfway across he fell into a deep channel. This, in itself, was not serious, but in his rucksack, in addition to the desperately needed tagging tool, was a car jack for the Land Rover. The combined weight of these two items, plus his belongings, took him to the bottom of the channel where he had to abandon his rucksack in order to avoid drowning. It lies there still. We now had three students with only one tagging tool.

Orty's arrival in January was equally astonishing. Mere enthusiasm for life never quite described the character of Orty, who seemed to have been born with a determination to exhaust about seven normal lives. During the previous season he had developed a passionate hatred of the road to Bhanga Nek and when we picked him up in Kwangwanase he was already three sheets to the wind. He said the only way to enjoy the trip was to drink his way through it. He also had the foresight to bring a long string of Chinese crackers, which he lit and hurled, as required, at the tribal dogs that normally gave the vehicle a warm welcome and a steady chase en route to the beach. That was the only trip I remember when, every 100 metres or so, the air was punctuated by a series of explosions.

The team began by establishing a patrolling system that remains in place today. Each student, accompanied by a Thonga assistant, would walk 8 kilometres of beach twice per night, one patrol going north and one south. The staff who undertook these demanding nightly foot patrols became known as 'turtle-trotters'. By standardising the sampling patrols, it has proved possible over the years to interpret changes in nesting densities with confidence.

It was also decided, for the first time, to split the student team, and two of us went down to live at Black Rock, a beautiful bay some 16 kilometres south of Bhanga Nek. In one fell swoop we had doubled the distance being patrolled on foot and, in addition, regular vehicle patrols now went as far south as Mabibi. Now 56 kilometres of beach were being regularly patrolled and the numbers of turtles of both species entering the record books started to increase, giving cause for much optimism in the Natal Parks Board.

We did, however, have serious problems with food and water supplies, which were both made more difficult as the road to Black Rock was, if anything, worse than the road to Bhanga Nek. When Cyclone Claude arrived in January 1966, our joyful isolation became much less pleasurable. The district was so badly flooded that no one could get to Kwangwanase and we started to run out of food. At Black Rock, John and I were beginning to envisage starving to death or, at least, having to eat raw fish, as matches were low and cooking oil was a thing of the past. Thanks to John blundering around the tent one night, the paraffin stove, along with our unique version of *vetkoek* simmering in our last pot of cooking oil, went over in slow motion. We watched in horror as the *vetkoek* were scattered and the oil soaked away into the sandy floor. It is hard to describe the anguish associated with these events, especially since we were in the middle of a cyclone that threatened to blow the tent away and had no prospect of replacing our oil or flour.

But all was not lost. When the rain stopped, John and I decided to patrol south together that night. Some 7 kilometres south of Black Rock we saw a light and what appeared to be a truck. Beach traffic was extremely limited in those days. Apart from the fact that all beach driving was banned, 4x4 vehicles were very expensive and a rarity, and trucks on the beach were unheard of. And so it remained. The 'truck' turned out to be a beached prawn trawler, the *Rocktail*, fresh out of Durban. At the height of Cyclone Claude's onslaught, the crew, thinking that they were many kilometres out to sea and being happy to ride out the storm, suddenly felt a solid thump and the ship stopped dead. Panic stricken, they launched their rubber lifeboat over the side and all leapt into it. It didn't move as it was in the shallow wash zone with only a few centimetres of water. The *Rocktail*, having been blown way off course by the gale, had been picked up on a gigantic wave and carried right up the beach and then dumped. It remains there to this day, now buried under several metres of sand, but living on in name as a local beach resort called Rocktail Bay Lodge, which is run successfully by Wilderness Safaris.

The light we had seen belonged to a local constable who had been stationed there to prevent the shipwreck from being plundered. The crew had walked to Kwangwanase earlier that morning and reported their plight to the police station. Happily, the constable was a trusting soul and John and I, by taking turns at distracting his attention, had soon stripped the boat of every egg, bottle of oil and anything else that was remotely edible. John made a bizarre impression as he gingerly climbed down the side of the *Rocktail*, clinging desperately to a trailing net with his south-wester pockets bulging with eggs. The *Rocktail* saved the day.

Over the next week or so, we steadily removed everything of value that we could lay our hands on. The booty included books, charts, tools and rope, but we found out too late that the difficult-to-access and flooded refrigerator had contained enough fresh meat to have permitted us to live like kings. Within ten days, the sea had smashed open the stern of the *Rocktail* and she broke up rapidly before any official salvage could take place and the beach gradually engulfed her remains.

The 1965/6 season was a wonderful experience for us students, cyclone and all, but there had been problems with feral dogs, some additional killing of adult turtles and extensive nest robbing. These problems were felt most keenly by the Natal Parks Board staff, who were intent on improving the security of the beaches.

Nick Steele, the warden of Umfolosi Game Reserve, having been told of the security difficulties, visited the beaches and wrote a report suggesting that horse patrols were the answer. Nick believed that horses were the answer to just about every problem and, with the approval of senior management, a major programme was unrolled to patrol the beaches on horseback. In November 1966 a herd of horses, which would have done Arizona proud, arrived with a large contingent of Zulu game guards from Umfolosi and other Zululand game reserves.

When I arrived with Mike Mentis and two other students in December 1966, our new officer-in-charge, Major Pat Temple, greeted us like the second coming. We soon found out why. Horses are splendid beasts – I had a string of four when I was a game ranger in the Drakensberg – but not on turtle beaches. Firstly, they required masses of water, which meant that we spent half of our time during the day ferrying water from the Malangeni to the main base near Black Rock where the horse teams had been stationed. As mentioned earlier, drawing water for ourselves had been a hated chore. This brilliant scheme more than quadrupled the task and we loathed it.

The horses didn't appear too enthusiastic about walking tens of kilometres of beach every day. During neap tides it was hard enough for us light-footed humans to struggle through the soft and yielding sand, let alone the hoofed horses, which rapidly lost condition. This, of course, called for the arrival at Kwangwanase of ever-increasing quantities of higher-quality horse feed and cubes, which necessitated even more trips to town. As we were still undertaking our nightly foot patrols and the one vehicle patrol, and because were always tired, we began to hate the horses.

To complicate matters, the Zulu horsemen were macho types and probably none too thrilled at their enforced exile to the wilds of Maputaland. This was not helped by the fact that Zulus traditionally, even as far back as the rule of Shaka, regarded the amaThonga as inferior and the practice was certainly continued on the beach. The whole sad affair ended with a trip to Kwangwanase for staff to buy rations and to collect yet another load of water at the Malangeni River. When we assembled the staff for our return, the Zulus had bought more than just food rations and one individual was somewhat belligerent, having enjoyed copious drafts of Lala wine in town.

When we stopped at the river, the Thonga staff July and Sonto, Mike and I started to transfer water from the river to the drums, using 25-litre plastic buckets. The Zulus made no attempt to help and amused themselves by passing snide remarks to one another. As most of the water was going to them and their damned horses, I felt that they should take their turn and approached the group, asking them to pitch in. The tipsy (and very large) fellow stood up over me and, in a belligerent tone, told me that this was work for labourers, not game guards, and he turned his back on me. I remonstrated with him and he raised his finger under my nose, asking whether I understood Zulu and then laughed, saying he would not take orders from a local. I knocked his hand away and told him in Zulu that I insisted that they help.

His immediate response was to pull out a knife and take a hefty swipe at my chest. Fortunately, thanks to the Lala wine, he was a little slow and I managed to pull back sufficiently to avoid the knife that just whiffed past my chest, which, at that instant, was as inverted as any chest could ever be. At this point, my natural sprinting talents came to the fore with a speed that surprised even me and I withdrew down the river as fast as my legs could carry me. While pausing for breath some 60 metres away, I was somewhat relieved to see the other guards leap on the man and wrestle the knife away from him.

The trip back to the beach, after things calmed down, was rather quiet. The net results of the little fracas were: Major Temple gave me a bawling out for knocking the man's hand away (apparently officers never strike a subordinate, even on the hand); a senior Natal Parks Board officer wanted me fired without bothering to hear my side of the story (God knows what the belligerent gentleman told him); and when I explained everything in detail, with the support of our Thonga staff, to Boy Hancock, the resident Natal Parks Board conservator of the region, his sole comment was, 'I would have shot the bastard! You will hear no more about it and the horse teams will be withdrawn.' We were friends for life.

The following season, we started to get down to the real business of researching turtles without the well-meant, but unfortunate, distractions that had faced us for the first few years. Even today I look back to the 1966/67 season when, had there been slight changes in scenario, my turtle career could have ended rather abruptly.

CHAPTER 5

# Tools of the Trade

*Part 1: The Humble Tag*

At the end of the 1965/6 season, Mike, John and I made our first attempt at writing a scientific paper. The first draft was not a great success, however. Bob Crass, the Natal Parks Board's senior scientific officer, gave our first draft about three out of ten, saying that he was amazed at how three intelligent young men could produce such a poor effort. He actually selected John as the victim for his sternest criticisms, saying that John's attempts at mathematical analyses were worrying. Bob's sharp guidance must have done some good because all three of us eventually earned PhDs and John was awarded the Order of Australia for his development of a software system for the Australian National Health Service.

We were suitably abashed, of course, and went back to the drawing board, eventually producing a paper that Bob proclaimed worth publishing, to our absolute delight. It was the second in a long line of scientific publications on turtles that appeared in the Natal Parks Board's scientific journal *The Lammergeyer*.

This paper, and many subsequent publications, would never have been possible without the information gathered through the use of the tag. Archie Carr, the spiritual leader of turtle conservation and research, told me in the beginning that we would learn nothing about

## Tools of the Trade – Part 1

adult turtles without the use of the tag – the most useful tool available to biologists. So, a marking programme was envisaged and right from the beginning the programme had its share of difficulties and disasters involving tags, which almost require a book of their own.

In the mid-twentieth century the tagging of marine creatures across the world was not even in its infancy. It hardly existed. In 1949 John Hendrickson and Tom Harrison had tagged some green turtles, *Chelonia mydas*, in Sarawak, and Archie Carr had started tagging greens with cattle ear tags in Tortuguero, Costa Rica. By the 1960s, Archie was convinced that the Monel metal cattle ear tag produced by the National Band and Tag Company, Kentucky, was the most suitable. Monel was a stainless nickel-copper alloy with a known resistance to corrosion and the National tag had an excellent locking device.

In 1963 such things were unknown to Hennie, John and Humph, but they were in no doubt that some form of recognition of individual turtles was necessary. The Natal Parks Board, however, who at that stage had only a very peripheral interest in sea turtles, considered metal tags an expensive luxury. So, Hennie, John and Humph were permitted to buy small plastic tags normally used for sheep. These tags stayed in place for a while, but were clearly not going to survive long years in the sea. Humph and John used these numbered tags on the trailing edge of the fore-flippers of both species, combining colours according to the locality along the coast where the turtles were encountered. From those pioneering efforts it was learned that some turtles could nest several times in a season, and the following year one or two were found coming back to nest. The study had begun in earnest.

By the time Mike and I arrived on the beach, the Natal Parks Board had generously purchased plastic cattle ear tags from a company called Dalton. This was a British tag and a considerable advance on the sheep tag. It was larger and consisted, like the sheep tag, of two halves, one of which had a coded numbering system and the other a return address and the word 'reward'. The tags were applied using an applicator rather like a pair of pliers. One half of the tag had a reinforced hole, and the other a pointed pin with an expanded head, which was forced through the flipper of the turtle and then through the hole where the expanded head locked securely. This held the tag snugly in place on either side of the flipper. It transpired that this tag was being employed on many other beaches. Today there are still one or two turtle programmes using Dalton ear tags for the same reason that we used them: their low cost.

We used Dalton tags for three years and, because they lasted better than the sheep tags, we soon benefited from their greater efficiency. But by 1968 we realised that the Dalton tags had their limitations. After a year or two, many returning animals had only a piece of the plastic pin embedded in the flipper, the two large 'wings' of the tag having broken off. Others had only a callus in the tag site, which suggested that the tag had been lost entirely. Loggerhead females obviously gave tags a hard time.

What is more, the tag site on the fore-flipper of leatherbacks proved to be very unsatisfactory for a number of reasons. Firstly, having such a soft skin, the tags could seriously wound the turtles during nesting as they vigorously swept sand to excavate the nest pit and to disguise the nest. The thumping of the Dalton tag against the carapace caused large scratches that could, on occasion, bleed prolifically.

The leatherback nesting swipe was not something to treat lightly. Many of us carried scabs on the shins from carelessly standing too close when a leatherback decided to sweep its fore-flippers. Humph was once knocked clean off his feet. The sand sent flying by the sweep was also dangerous and could carry for up to 10 metres. If you were too close, a flipperful of sand, driven like a bullet into one's eyes, was a memorable experience.

We also found that many tags on leatherbacks, which had been in place for only a short period of time, were already cutting their way through to the edge of the flipper. It was clear that the massively powerful swimming strokes of the turtles were causing the tag to erode its way through the flesh and would soon be lost. This problem was addressed by moving the tag site on leatherbacks to the hind-flipper under the projecting carapace. Our programme was the first to introduce this innovation.

A change, in 1969, to the large, bright yellow, circular (and free) plastic tags used by ORI in Durban for tagging sharks, proved an

## Tools of the Trade – Part 1

unmitigated disaster. Firstly, using a leather punch, one had to punch a hole through the flipper before inserting the pin and forcing the female half over the pin by hand to lock it in place. This process took some time and by the end of the season we were already finding an unacceptable number of lost tags because the flesh where the hole had been punched was becoming infected. The swelling of the flesh was clearly enough to pop the two halves apart. There had to be a better way and Archie kept on telling me to try the Monel tag.

In 1969 I managed to get funding from the newly established South African Wildlife Foundation (the South African appeal of the World Wildlife Fund) and I ordered the first tranche of Monel tags from the National Band and Tag Company. The first series was stamped with an 'A' prefix followed by a number on the upper side and a return address and 'reward' on the lower side. The age of metal tags had begun in Maputaland.

For the next 35 years it was the intention to start every season with a change of prefix letter, always followed by numbers. It would, therefore, be possible to identify the year of tagging almost instantly. I say almost, because, on one or two occasions, ordered tags did not arrive punctually and the season would start, frustratingly, by our having to use leftover tags from previous seasons. The turtle itself remained individually recognisable from the number, of course.

The Monel tag was a huge improvement and data concerning turtle activity, and migrations began to accumulate rapidly, but, contrary to expectations, there were still problems with tag loss. Although we had changed the tag site on leatherbacks from the fore-flipper to the hind-flipper, which greatly improved the survival of the tags, there were still worrying losses in leatherbacks – and loggerhead losses were also significant. We also started to retrieve tags that showed quite clearly that Monel was not as corrosion resistant as we had thought. Some tags were found almost completely corroded away, although still in place, with the tag numbers having gone and the locking mechanism failing. The National Band and Tag Company had been aware of this problem for some time and had started to produce a tag made from Inconel, a nickel-chrome alloy that was more corrosion resistant. It was unfortunately only available in an unsuitably small size and was never used in our turtle programmes.

We were not the only scientists beginning to worry about this problem. My friend and colleague Colin Limpus from the Queensland Department of the Environment was already experimenting with double tagging and was encouraging a company called Stockbrands in Western Australia to research a better tag. Stockbrands eventually produced a very expensive, but totally corrosion-proof, tag made from titanium. In 1984 we purchased our first titanium tags and these, with an improved locking system, once again increased our tag returns and have been used annually ever since.

Titanium tags did, however, prove that there is no such thing as a perfect tag and, especially during the early years, they were found to be somewhat more brittle than Monel and quite a few snapped as one applied pressure with the applicator. This was worrying. But then came the advent of the PIT tag, which is widely used today in the pet trade and in horse husbandry. The PIT tag (or Passive Internal Transponder) is a small tag, 11 millimetres in length, which is injected under the skin of an animal and read electronically with a special reader. In 1991 we started to double tag about 100 turtles a year, with the external titanium tag in the flipper and a PIT tag buried in the flesh of the shoulder. The cost of about R20 per PIT tag placed an additional strain on the budget for the programme and this necessitated the limit of 100 per season. Recovery of one or the other would identify the individual turtle and provide information on the efficiency of the tags.

Over the years all of these tags have produced results that astounded us.

As mentioned previously, size is no indicator of age, but we discovered that once the roughly 1-metre long loggerhead turtle has nested, it virtually stops growing. We believe that most of the growth takes place as it is maturing and if the animal has grown up in favourable conditions, with warm water and adequate supplies of food, it will probably grow to be a large turtle. If it has survived through more marginal conditions, it will probably be a smaller animal on arrival at the nesting beaches. Some loggerheads, such as those of the north Atlantic Ocean, spend much longer at sea, travelling over a much larger area of ocean, and tend to be, on average, larger than our Maputaland females.

Modestly sized though they may be, loggerhead turtles nesting on the Maputaland coast are not just local animals. Tag returns over the years have shown that they are drawn from the entire east coast of Africa from Cape Agulhas to Somalia and from the waters all around Madagascar to the Seychelles and Mascarene islands, which they are known to visit as adult migrants. Some even go round the Cape of Good Hope and live on the west coast of South Africa. This means that our loggerheads are distributed along some 5,000 kilometres of African coastline and our Maputaland nesting population is drawn from some 15,000,000 square kilometres of the Indian and Atlantic oceans – a massive area.

Loggerhead females, having completed a nesting season and having lost about 25 per cent of their body weight in the process, return home with real determination. Recovered tags have revealed that the turtles are capable of swimming nearly 40 kilometres per day, every day, for up to two months, in a non-stop return to their feeding grounds. They may remain there for years before returning once again to their nesting beaches, where they were born some 24 years before.

Tag recoveries have shown that many loggerhead females appear to visit the nesting beaches only once in their lifetimes. Others have returned in two, three, four, five and up to seven separate seasons to lay their roughly 400 eggs in the summer months between October and February. Nesting in consecutive seasons is quite uncommon and the most common intervals between nesting seasons are two to three years. Of course, there is always the odd one out. One female was seen nesting at a spot on the beach known as 8N. She was found only once during the season and she then disappeared for 16 years, before being found again nesting only once during the season at 8N. What is more, she had a Monel tag that looked as if it had been put on the night before.

Incidentally, one often reads in the popular press that turtles return to nest in the same spot, whether within a season or in different seasons. This, alas, is not borne out by results, which should not in any way detract from an appreciation of the amazing navigational ability of sea turtles (see Chapter 14).

Early publications on sea-turtle nesting, based on rather limited tag returns by re-migrants, concluded that sea turtles had nesting cycles that were regularly repeated at either two- or three-year intervals. Later studies demonstrated that there are certainly females that were capable of this, but many more did not reflect such regular cyclic re-migrations. In Maputaland, we were the first to report that leatherbacks varied their re-migration intervals. Some females would come back after two years and then again after three years and vice versa. Over the years, we have recorded many variations in re-migration intervals and it is clear that the ability of the turtle to withstand the rigours of a nesting season dictates a re-migration interval rather than an instinctive commitment to a regular cycle.

Many leatherback females in Maputaland lay over 1,000 eggs in a single season and have returned over the decades after intervals of two, three four and five years. They vary their re-migration intervals far more frequently than loggerheads and regular and irregular nesting cycles occur almost equally.

To my surprise one year, right at the beginning of the nesting season, a leatherback female carrying a metal tag was observed nesting on the beach at the Storms River mouth in the Western Cape. The beach was over 1,000 kilometres south of Maputaland. This news received considerable attention because records of any sea turtle nesting far away from her natal beach were, and remain, exceedingly rare. None of us could understand what she was doing nesting that far south on a cold beach.

The riddle was solved when we started to use satellite tags.

CHAPTER 6

# Tools of the Trade

*Part 2: The Satellite Tag*

Many real surprises came with the advent of satellite tagging. Not only were our theories deduced from metal-tag recoveries regarding the movements and behaviour of loggerheads confirmed, but we also learned what the leatherbacks did after nesting, which had been a bit of a mystery to us – we had recorded only three long-distance recoveries of metal tags in nearly 30 years. What was discovered about leatherback post-nesting movements turned out to be simply amazing. The term 'pelagic animal' took on a whole new meaning.

In 1995 Prof. Floriano Papi from the University of Pisa in Italy, contacted us and asked whether we would be interested in a cooperative study following the movements of sea turtles fitted with satellite transponders. I am not sure whether sound can travel all the way to Italy, but the howl of delight that I gave must have nearly made it. At a minimum of US$5,000 (about R25,000) for each satellite tag, we had never even been able to contemplate the use of this wonderful tool – almost the ultimate tag available for biologists. I might add that the running costs of the tag are also on the steep side. A daily download of information via the satellite costs US$50 per tag. You can imagine the enthusiasm with which I wrote back to him. As far as I remember, over half the letter contained the words 'welcome' and 'yes' in more or less equal quantities.

So, in December 1995, armed with seven satellite tags, Prof. Papi and his colleague Dr Paolo Luschi came to Bhanga Nek for the first time. This was a thrilling time for us as our team helped Floriano and Paolo prepare the loggerheads in order to glue (using dental adhesive) the satellite transponders to the upper surface of the carapace. The site had to be clean and dry before the glue was applied and the tag firmly embedded at such an angle that the aerial would break the surface when the turtle rose to breathe. When it broke the surface, the position of the turtle was transmitted to the Argos satellite and later downloaded and the collective information reported to Floriano and Paolo in Pisa.

The results were satisfying indeed. The loggerheads, having completed their nesting season, all headed north along the coast of Mozambique, following routes that we had surmised from tag recoveries over 30 years (although the satellite routes were somewhat further out to sea than we had imagined). Following that first season, many more satellite tags have been fitted to loggerheads – thanks to the Italian team (expanded one season to include Drs Resi Mendacci and Allesandro Sale), Dr Graeme Hays of Swansea University and the South African team of Dr Ronel Nel of the Nelson Mandela Metropolitan University, Mike Meyer and Herman Oosthuizen of the South African Department of the Environment, Marine and Coastal Management Branch. This past season (2011/2012) has seen more tagged loggerhead females following migration routes north and south of Maputaland and across the Mozambique Channel to Madagascar.

What has been of particular value is the proof, derived from hundreds of satellite records, that loggerhead turtles do indeed have a specific feeding territory to which they return after a nesting season. It would appear that, once established, this home territory is used for the rest of their lives. Once a turtle has established a feeding territory, we believe that it defends it vigorously. In the main tank of the Durban Aquarium, for example, loggerheads were seen to become quite vicious towards a newly introduced loggerhead. On one rare occasion, when four loggerheads were placed in a modest-sized tank, three of them attacked the fourth and killed it.

These dramatic territory records may explain why adult loggerheads are distributed so widely over the western Indian Ocean. Tag recoveries have shown that some are resident around the Cape in the Atlantic, up the west coast. One female currently being tracked has shown that they are relatively impervious to cooler waters as she is currently settled some 133 kilometres south of Cape Agulhas and only 61 kilometres from the edge of the Agulhas Bank. Although, thanks to the warm and pulsing Agulhas Current, the temperature in the area is at present (February 2012) a pleasant 21° Celsius, during the winter months the temperature will drop to 14°–15° Celsius, and lower, as the Agulhas Current weakens.

After their epic journey as hatchlings and juveniles around the southern Indian Ocean gyres, sub-adult loggerheads are brought back by the currents to the mainland coasts of Africa and Madagascar. At this stage, they have attained a carapace length of approximately 60 centimetres. They then change their lifestyles, giving up their existence of floating near the surface, which they have led for a decade or more, and begin diving for food. This probably brings them into conflict with adult loggerheads that have long-established feeding territories and are ready to defend their precious food resources. The resident adults drive the youngsters away and, thus rejected, the sub-adults move on, from site to site, until they eventually find an unoccupied site. If the site is suitable, the growing turtle will settle and remain until mature and capable of enduring the demands of a nesting migration. At this point, both males and females will depart for their birthplace to contribute to future generations.

Additional records, derived from the random capture of loggerhead turtles in the widespread and efficient set of shark nets along the coast of KwaZulu-Natal, have shown that the majority of loggerhead turtles caught are sub-adult. This suggests that the sub-adults explore more and are much more vulnerable, being caught throughout the year. Adult and tagged loggerheads tend to be caught either at the beginning or the end of the nesting season, when they are going to, or returning from, the nesting beaches.

It is a sobering thought that some sub-adult loggerheads may have to wander for years, over thousands of square kilometres of suitable feeding areas, before finding an empty site – either vacated by a turtle that may have died of natural causes or perished in the nets of a passing prawn trawler.

When the Italian team released the first satellite-tagged leatherback, however, surprise after surprise enthralled us as the flow of locality records was received.

As mentioned, we had had only three metal-tag recoveries of leatherbacks over a 30-year period and absolutely nothing could be deduced from them other than the fact that leatherbacks actually wandered near those sites. One was killed by a trawler off Beira, another washed ashore on Ile Saint Marie on the east coast of Madagascar and a third was killed in the shark nets off Warner Beach. This suggested that leatherbacks, after nesting in Maputaland, could set off in at least three directions. The how, where and when of the movements remained unanswered.

Over the next 15 years, satellite tags gave us the answers. We were overawed by the very first one in 1995, and this set a trend of growing amazement at what these magnificent wild animals could achieve.

Leatherbacks are very big animals, reaching nearly 3 metres in length and the very largest probably exceeding a ton in weight. The heaviest ever weighed, a casual visitor to Wales, was 916 kilograms and lengthwise, by our local standards, a fairly modest-sized individual. Once committed to an action, leatherbacks are almost totally dedicated to achieving it. This explains why, during the nesting season, so-called 'exploratory or non-nesting crawls', very common in loggerhead behaviour, are so rare among leatherback emergence records. They are not easy to disturb once they have started the nesting process and, even if they have recently emerged from the sea, some dedication is required in order to deflect them from their purpose. So, whatever one wants to do with leatherbacks is a major and reasonably hazardous endeavour. Tagging leatherbacks with simple metal tags requires care; fitting them with satellite tags requires fortitude.

Although this has been done with special glues recently developed by some American biologists, in the early days, satellite tags were never attached directly to the skin of the carapace. In Maputaland the idea never found favour because, as the name implies, the carapace of the leatherback is covered with a soft, smooth, leathery and, incidentally, prettily marked skin. It is not particularly tough and bleeds easily if damaged. As a result, all over the world, harnesses have been designed that incorporate a base plate to which a satellite transponder can be affixed without touching or damaging the skin. The harness and the plate should not hamper the swimming movements of the animal once freed.

Floriano, Paolo and our team applied ourselves to the task of making local harnesses after receiving guidance from Dr Scott Eckart from California, who pioneered the technique. The materials had to be durable and the harness had to be adjustable in order to fit any female with a carapace length from 140 to 206 centimetres. The difference in bulk over that range is enormous. Although not a work of art, the first harness, built like a rucksack, but using extendable rubber linking, together with hard plastic clasps (the same as those used to ensure that luggage does not burst open if the locks fail), was successfully assembled. We gathered round a nesting leatherback with some trepidation, as we expected some difficulties in fitting it. How right we were.

No leatherback worth her salt offers you one second's worth of cooperation. She arrives with a purpose and the team's interests are of no consequence to her. You cannot lift a leatherback female and neither can you stop her from moving either her flippers or her body. Timing, really outstanding timing, became the key to success in fitting the harnesses. Anything short of that resulted in grazed fingers, bruised legs and the occasional open wound. Why you may ask? Because the harness had to be manoeuvred under her as she walked back to the sea and the closer she got to the sea the faster she moved. Finding the bits to connect, making the connections and checking to see whether they were secure (in addition to the liberal use of cable ties to hold the clasps together, the ends of which had to be carefully clipped) became more frantic as she closed with the sea. The entire process was accompanied by a growing crescendo of cries of encouragement, pain and frustration from a variety of team members, each experiencing his, or her, own particular emotion or animus. An excited and slightly wounded Italian, with his hand caught under a leatherback flipper, is a sight worth seeing.

It was not without a sigh of relief, mixed with a glow of triumph, that we watched our first leatherback disappear into the waves with the satellite transponder securely in place. The effort was totally worthwhile because, over the next few months, she travelled 7,000 kilometres, sweeping down the eastern edge of South Africa in the Agulhas Current, turning south near Cape Agulhas and heading for several hundreds of kilometres deep into the southern Indian Ocean. She very nearly hit the Subtropical Convergence before turning east and heading for Australia. The water is cold down there, but she showed no immediate reaction to encountering the very cold water so far south. She never appeared to stop and could cover well over 130 kilometres per day. We were very excited by these movements and were desperately keen to see how far she would go, but then, as is the case with so many satellite-tagging ventures, the battery failed and she disappeared. Disappointment hardly describes the feeling of losing a turtle after months of daily tracking and this is a problem that has yet to be solved.

In subsequent years, information gathered from satellite tags has made it clear that leatherbacks, quite unlike the other species of sea turtles, regard the world as their oyster. Disavowing the coasts, but becoming truly pelagic, they do not adhere to any set pattern of migration, but instead go where the food is to be found. From Maputaland, several females have gone north, meandering up the Mozambique Channel as far as the Comoros. Others have swum eastwards, in one case 1,000 kilometres past Madagascar to beyond Rodrigues Island and then north up to the Chagos Islands. A recently tagged female has travelled south-east to the Madagascar Ridge.

The most dramatic and frequent voyages, however, have seen leatherbacks passing the Cape of Good Hope, hundreds of kilometres south of the African land mass, then turning and swimming into the Atlantic Ocean and northwards to Namibian and Angolan waters. One female made a 20,000-kilometre swim in ten months; our knowledge of her travels ended when she was in the deep Atlantic, 1,000 kilometres west of Luanda, Angola. We lost her when, as with all the other satellite tags we have used, the batteries gave out.

The satellite tracking also solved the mystery of the leatherback nesting on the beach at the Storms River mouth. She was clearly a female that had spent a few years in the Atlantic and was, somewhat tardily, making her way back to Maputaland around the Western Cape. Her first clutch of eggs obviously matured earlier than expected. It was a sort of leatherback 'baby in the taxi' event. Her next clutch was laid on a Maputaland beach over 1,000 kilometres further north, as was more normal for her.

If one can be impressed by this feat, consider what else leatherbacks can do. In 1998 some of the satellite tags were more sophisticated, having the ability to record temperatures, direction and depths of dives, presumably for feeding. For the most part, diving 50 metres or so is a common activity for leatherbacks, but every now and again they show their real diving skills. In South African waters the deepest dive, recorded in 2002, is 940 metres, which is an impressive depth by any standards, putting the leatherback in the same diving class as the sperm whale. Colleagues in other parts of the world have recorded leatherback dives of well over a kilometre, so such behaviour is not uncommon.

Another amazing adaptation is their ability to resist the pressure at such depths. Unlike all the other sea turtles, the carapace and plastron of the leatherback are not solid, but made up of thousands of small bones loosely sutured together. It is thought that, as the turtle dives, the small bones come together, closing the sutures until the bony carapace becomes almost as solid as those of other turtles, forming a compact box in which the sensitive organs are protected.

Adding to the wonder of the leatherback is its ability, as we have seen through satellite tagging, to enter and feed in extremely cold water. At 700 kilometres south of the Cape, temperatures drop below 10° Celsius in the winter and leatherbacks have been seen among the icebergs south of Greenland in the Atlantic. What makes this behaviour possible is the leatherback's ability to maintain a body temperature up to 20° Celsius higher than ambient temperature. In other words, like mammals, a leatherback maintains a steady internal temperature almost irrespective of the temperature of the sea water around it.

Altogether the leatherback turtle is an amazing creature.

A new research tool has recently become available. It is called a data logger and has the ability to record the movements of turtles at almost any stage of their lives. A Japanese colleague, Dr Tomoko Narazaki of Tokyo University, has already fitted them to Japanese loggerheads during the off-nesting season and has shown that the turtles clearly reorient themselves when they break the surface of the sea to breathe.

We are currently faced with two problems. The first serious problem being that one has to recover the data logger before the data can be extracted from it. Tomoko has designed a release system and has fitted the logger with a radio beacon, which will hopefully make it easier to retrieve when it floats to the surface. As these loggers are small, the recovery is not that simple in rough seas and would be very difficult along the high-energy coast of Maputaland.

As yet, we have not had an opportunity to try any data loggers in South African waters, so there is still much excitement to be had. This tag may provide us with information that will confirm the theories of how sea turtles navigate their way across thousands of kilometres of open ocean.

CHAPTER 7

# The Search for the Lost Decade

## Part 1: The Nuclear Threat

Everyone knew that turtle hatchlings entered the sea after emerging from the nest, but precisely what they did and where they went was an intriguing problem facing turtle biologists in the late 1960s. Prof. Archie Carr coined the phrase the 'lost year'. Archie was primarily involved with green turtles in the Caribbean and he had established the resident habitats of many year-old greens (Archie famously referred to these as 'chicken turtles'), but was in pursuit of information regarding the first year of life. His interest was, therefore, much more limited than researchers of loggerheads and leatherbacks, for which species, in our opinions, he was proposing far too short a time of mystery. We felt that the 'lost decade' would be a more appropriate name for the period of a loggerhead's life cycle of which we knew nothing.

One of the problems in dealing with turtles immediately after they emerge on to the beach is that they are very small. At only 44–50 millimetres in length, attaching a metal tag, the normal marker used on adults, is simply not an option. Any small tag would soon be outgrown and lost; any larger metal tag would cause the hatchling to sink like a stone. Both would almost certainly handicap the animal as it entered the sea and make it more vulnerable to the startling array of predators that awaited it there. Regardless of these concerns, there were simply no small metal tags available on the market and we could not afford the larger Monel tags used by Archie and his colleagues in the Americas.

As a result of these difficulties, for the first few years in Maputaland, the turtle teams made no attempt to track hatchlings, but the question had been posed and intrigued us no end. So, in 1967 we decided to bite the bullet and make an attempt to follow our hatchlings.

Leatherback hatchlings were ruled out as experimental subjects for the simple reason that they were very much in the numerical minority on our beaches. The annual numbers of nesting turtles were made up of about 10 per cent leatherbacks and 90 per cent loggerheads. Leatherbacks were, therefore, far more precious and, because they had a soft shell to which nothing could be attached without exposing the animal to fatal infection, it was decided that we would never try to track leatherback hatchlings.

It fell to the delightful little loggerhead hatchlings to serve as an indicator species. We focused our attention on marking them so that we could find out where they went, how much they grew in the wild and, this was the dream of dreams, how long it took them to reach adulthood and whether they came back to their natal beaches as adults. What appeared to be a rather modest set of objectives took us some 40 years to achieve with certainty and involved the catching and marking of nearly 350,000 hatchlings.

Of every 1,000 loggerhead hatchlings that enter the sea, only about two reach nesting maturity. In addition to this, although the sex ratio of loggerheads at maturity is roughly one male to one female, the males never emerge onto the beaches unless they are dying or dead. So, we had to believe, against all probable odds, that we would have the good fortune of finding, among the multitude of nesting adult females, the odd survivor of the hundreds of thousands that had been marked. This was not a programme to be entered into by a pessimist.

In 1967 we decided that we would start tagging hatchlings. Discussions with Bob Crass, the Natal Parks Board fisheries biologist, had been fruitful. He confirmed that the attainment of our four objectives was necessary to understand the life cycle of loggerheads and, hopefully, make sound management decisions concerning turtle populations. After lengthy discussions on what was available, Bob suggested small numbered plastic tags that he had deployed successfully on freshwater fish. They were about 10 millimetres long, bright red in colour and normally injected into the body cavity of a fish. This was clearly not the ideal tag as recovery depended on people walking on beaches and noticing the tag on a stranded animal. As we had nothing else on offer we accepted 1,000 tags and set off to the beaches, wondering how we were going to utilise them.

By the time we got to Bhanga Nek, we had decided that we would drill two small holes on the side of the hatchling's carapace, two matching holes at the end of the tag and then insert a small piece of fishing nylon through the matching holes. The exposed ends of the nylon would then be heated and melted, thus holding the tag firmly on the underside of the carapace. By pre-preparing lengths of nylon, we anticipated that we would be able to deal with a hatchling in a couple of minutes. In fact, even after becoming relatively skilled at this process, it would take us more than a couple of minutes to deal satisfactorily with a small, vigorous and agitated hatchling.

Our strategy didn't take into account the fact that once hatchlings emerge onto the beach they are driven by an urge to survive that embraces a frantic, continuous four or five days of swimming, from which nothing will deter them. The rather delicate task of drilling holes through the edge of the shell was not assisted, nor appreciated I am sure, by the struggling hatchlings. The necessity of bringing heated wire near the carapace to melt the nylon, which was an improvement on the naked-flame we used initially, did not endear the technique to them either. We were lucky to finish tagging a hatchling in a little under five minutes.

During the course of the 1967/8 season we succeeded in tagging and releasing some 623 loggerhead hatchlings – not one of which was ever seen again.

After all that labour we were, of course, disappointed, but not surprised. The most we expected from the tags was a lifespan of two months before the growth of the animal forced it to drop off. A few tags on captive hatchlings stayed in place in excess of two months, but their growth rates were very poor and we did not regard the results as an indicator of what one could expect in the wild. One must bear in mind that hatchlings of any turtle species are under pressure to grow rapidly, as every centimetre gained in length reduces the range of predators that can eat them.

It was time to think again and, as preparations were underway for the next season, we began to think nuclear. Cobalt-60 is a radioactive element available in wire of varying thicknesses. It has a short half-life of 5.27 years and would thus be available as a source of information for probably as long as a turtle survived. The basic plan was to inject a 2-millimetre piece of cobalt-60 wire under the numerous conveniently placed carapace scutes (scales covering the bony carapace) of a loggerhead. Each year we would choose a different scute under which to place the wire. This would enable us, with the aid of a portable Geiger counter, to ascertain the age of a marked individual and whether the turtle came from Maputaland.

The concept gained general approval from the authorities at the Natal Parks Board and the Oceanographic Research Institute, to which I had moved after the 1968/9 season to commence my postgraduate studies. An application to the Atomic Energy Board in Pelindaba received cautious, but courteous, support and I was issued some samples of cobalt-60 wire of the appropriate thickness, with a proviso. The institute (me) was expected to carry out experimental tagging of captive hatchlings to ascertain the risk of tissue damage for a minimum of eight months after the insertion of the wire. Following a positive report on that research, the Atomic Energy Board would consider issuing the necessary permits. The downside was that we had to wear exposure badges every time we went near the experimental tank and had to submit these badges to the Atomic Energy Board at regular intervals.

That put to bed any hope of marking hatchlings in the 1968/9 season. At the end of a year of experimentation with the cobalt-60 wire, no trace of damage was visible under microscopic examination and a positive report was submitted to the Atomic Energy Board with a request that I be permitted to take the technique into the field. It really was very promising, as the wire was easily and unerringly picked up by the counter and gave a very sharp and accurate location of the site of the tag, which was absolutely ideal for our purposes.

Ah, but 'the best-laid schemes o' mice an' men / Gang aft agley' and the Atomic Energy Board turned down the application. They insisted that it was too dangerous, quoting, as a terrifying possibility, that some young boys might find the radioactive hatchlings on the beach and place them in their pockets, thus irretrievably damaging the boys' reproductive organs and affecting their future procreation activities. This struck me as appallingly sexist, since they didn't even consider the possibility of a young girl popping a hatchling into her panties. As I have always regarded the mega-surplus of human beings as the real threat to the survival of our biodiversity (and I am still of this opinion), I complained bitterly to my director Dr Allan Heydorn, asking why the fools could not see that this was a distinct advantage. Allan, although sympathetic, felt that appeals would achieve nothing, so the great nuclear threat to South Africa's children died stillborn.

So, yet another season went by without any progress in the hatchling-marking saga, but the wire brainchild refused to go away. The following season, in 1969/70, saw the resurrection of a slightly modified plan. We would use stainless-steel wire rather than cobalt-60 wire, which might have caused permanent damage to the turtles, and X-ray a hatchling or sub-adult to discover the presence of the wire tag. Laboratory tests on injected wires on young turtles, right up to sub-adult, proved that the method worked. Granted, the new plan was neither as elegant nor as easily managed as the radioactive wire (we immediately abandoned any possibility of getting an X-ray machine onto the beaches during the nesting season and manhandling adult females into it), but it was felt that, although more limited, it had possibilities. So, we arrived with plenty of wire, syringes and alcohol (both for sterilisation of the wire and syringe needles, and to help while away the hours).

The technique was a huge improvement on the old freshwater fish tags and we could mark hundreds of hatchlings in an evening with minimal discomfort to the young turtles. We were delighted that we had inserted steel tags into, and released, 5,000 loggerhead hatchlings by the end of the season. A new spirit of optimism swept the turtle team as we broadcast news of this innovation through the radio and the printed media from Natal to Cape Town, asking our informants and volunteers along the coast to search for and collect any stranded hatchlings. This is where the problems began. Over the next several months, hatchlings arrived in boxes by the dozen and a new challenge was born.

In those days, the cost of X-rays was high (not that things have changed a great deal) and suddenly I was confronted with the need to X-ray many dozens of hatchlings and, after making enquiries about costs and availability, I realised that my budget, slender as it was, could not afford a lengthy series of X-rays at commercial rates. Driven by near desperation, I approached the formidable figure of the senior sister-in-charge of the radiology department at Addington Hospital, which was not far from ORI. I had been warned that she was a strict disciplinarian and I was somewhat nervous when I approached her for help.

To my surprise she gave me a sympathetic hearing and immediately produced a free solution to my problem. 'Young man,' she said, 'we have an endless line of patients coming to the hospital for large X-rays and all you have to do is be available to come at short notice with your bag of specimens (she had first asked with some apprehension how large my turtle specimens were likely to be) and we will simply spread them around the available space on the X-ray plate.' She was clearly delighted at my response and thereafter, for a few months, I would receive a call from the radiology department to hotfoot it over there with however many hatchlings I had awaiting X-ray. This worked brilliantly. Most patients showed great interest in the tagging programme and, although we never actually found a turtle that had been implanted with a piece of wire, the anticipation of eventually finding one was high. In any case, the array of oddly shaped X-ray cuttings in my office proved to be a constant source of wonder and discussion.

It all ended abruptly, however. One morning the clarion call was received and I bolted to the hospital with a bag, perhaps more full of specimens than normal, and rushed into the X-ray room. Without further ado, I started to distribute the hatchlings along all visible the edges of the X-ray plate. Alas, this was not quite enough to cater for all the hatchlings because the patient, a lady of generous proportions, had distributed herself over most of the plate. Showing some enterprise, I thought later, I sought available nooks and crannies to squeeze in the shortfall and by accident tweaked the lady's body with a hatchling. Not, in retrospect, a life-threatening action.

The lady watched me with growing alarm, as the X-ray technician had neglected to make a formal introduction. So, I must put her reaction down to being poorly informed. She erupted (I believe that is the appropriate description) and started to scream blue murder, scattering my hatchlings all over the floor as she writhed in anguish. She was apparently not an animal lover and immediately threatened to sue the hospital, her voice reaching a decibel level that soon had the sister-in-charge hurtling into the room. Naturally, I fled and after the general hubbub had died down the sister-in-charge emerged. She informed me, clearly with some regret, that what we had been doing was stretching the rules a bit and would, in order to prevent Addington Hospital from being the subject of lurid headlines and costly litigation, have to cease forthwith.

The final blow to the stainless-steel wire experiment came when an enthusiastic, but misguided volunteer in East London took about ten hatchlings to an X-ray clinic, which carefully X-rayed each one separately, and then sent the plates to me with a bill that took my breath away. It took several letters and an appeal from my director in writing to persuade the clinic to waive the charges. All this excitement convinced Allan that the technique had serious shortcomings and that I should look for another method with more certainty of success.

And this I did.

CHAPTER 8

# The Search for the Lost Decade

*Part 2: 'Isn't that one of your f---ing marked turtles?'*

Colin Sapsford, one of the brightest lecturers in biology at the University of Natal, Durban, had long been interested in matters physiological and the Maputaland sea turtles had attracted his attention. He was also a good friend, with a hysterical sense of humour, who could be blunt to the point of rudeness, when appropriate.

Colin came up to the turtle beaches to study the internal body temperatures of nesting turtles and accompanied me on a number of routine tagging patrols. On every outing, Colin was armed with a long black rubber tube at the end of which, encased in silicone, was a temperature-reading thermistor. When encountering a turtle, this tube was heavily smeared with Vaseline and plunged deeply up the cloaca of the unfortunate turtle. (Shall we say that the turtles did not display any great pleasure at the procedure and after a week I had the distinct impression that they were hunching up at our approach. In addition, trying to explain to the appalled local staff what was happening proved a challenge. Science is not always an easy road.) Some time was allowed for the temperature to stabilise and the readings of core temperatures were taken, proving that the Maputaland leatherbacks could, like those in other parts of the world, maintain their internal body temperatures. We recorded in January that the Maputaland leatherbacks had body temperatures 4° Celsius above sea temperature. This is a remarkable achievement for a reptile and a clear explanation of why leatherbacks have such a regular ten-day inter-nesting interval. Loggerheads, however, cannot and their temperature and activities follow the dictates of the ambient temperature. When the sea temperatures are low, loggerheads take longer to mature their next clutch of eggs, so that at the beginning of the season the inter-nesting interval may be 21 days and by late summer, with higher sea temperatures, the inter-nesting interval can drop to 13 days.

Colin and I spent many hours discussing techniques used on the programme and I had given him some light entertainment describing the various spectacular failures we had experienced while trying to find a way of marking hatchlings for future recognition. He was very encouraging, while remarking that he was a bit of a specialist in spectacular failures himself.

In 1971 I attended the second meeting of the IUCN Marine Turtle Specialist Group in Morges, Switzerland, and this matter was discussed at length with a number of my colleagues. Dr Bob Bustard, then working in Queensland, casually remarked that he had tried notching marginal scutes on hatchlings from Heron Island and the technique seemed to work. Still smarting from my wire experiments, I noted this and resolved to consider the technique on my return.

*Notched loggerhead hatchlings*

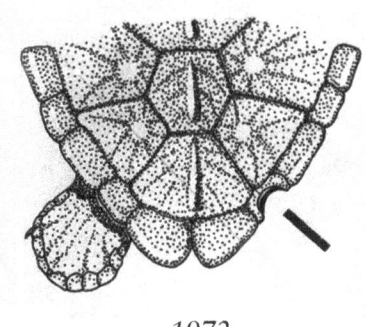
1972

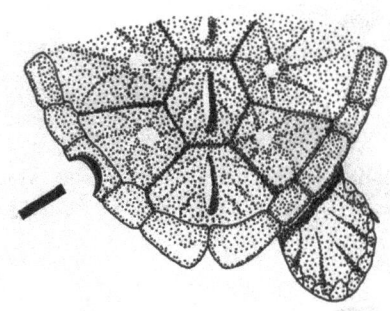
1973

Allan Heydorn felt that it was worth a try and, although it involved some mutilation of the hatchlings, we felt that this was at least more ethically responsible than the use of radioactive wire, which might have permanently affected the reproductive health of the turtles chosen for implants. (In our defence, the hatchlings that were to be injected with radioactive wire represented a miniscule proportion of the annual production and, at the time, this had seemed a reasonable experiment.) So, in the 1971/2 season, we started notching loggerhead hatchlings.

Notching is a technique eminently suitable for these hatchlings. Each hatchling comes equipped, normally, with 24 small marginal scutes evenly distributed around the edge of the carapace. Using a common, easily available (and inexpensive) leather punch, it was a simple matter to excise a selected scute and its underlying carapace, which left a neat hole where most of the scute had been. Our first efforts improved rapidly as we discovered that by holding the hatchling upside down it was possible to ensure the notching did not break any blood vessels, thus leaving the excision clean and bloodless.

The basic theory was that, as the carapace underlying the scute at this stage consisted primarily of cartilage, the edges would heal once the piece was removed and a permanent indentation would be visible on the edge of the carapace. By carefully sterilising the tool and treating the wound with gentian violet (recommended by a leading physician, it was applied by soaking a pipe cleaner with the solution and drawing it through the notch), it was hoped that little harm would come to the animal. The excision would most certainly not affect the swimming of the animals at all.

The process proved to be very rewarding. Hatchlings were collected in the second half of the season, opportunistically by foot patrols and more efficiently by driving, with one of the team scanning the upper reaches of the beach using a strong headlamp. We would look for the bright white line left in the dry upper beach sand by the small horde of 100 or so hatchlings heading for the sea. In time, some of the staff became so skilled at picking up such signs that it required the track of only one early emerging hatchling to alert us to an available clutch. All hatchlings available on the beach (we were occasionally too late to get an entire clutch and often too late to catch any) were then collected, including those attracted back on to the beach by shining the headlamp into the sea. Hatchlings and adults respond very positively to strong lights and this characteristic enabled us to maximise our collections.

In the beginning, individual clutches were held in separate containers so that the clutches could be independently measured and weighed, but in later years all hatchlings went into a collective plastic container in the vehicle. Foot patrols brought clutches back in empty orange pockets, which ensured that they had plenty of air. Once back in the camp, the research team, having returned from their 16-kilometre walk, would sit up until the early hours notching and treating the hatchlings with gentian violet. This sometimes took four or five hours and, if we had been particularly successful and caught over 500 or more hatchlings, it was all hands on deck, and students, staff and even guests were roped in to notch and treat the tiny turtles. The aim was to get the job done as quickly as possible and get the hatchlings into the sea, preferably well before daylight. As mentioned earlier, hatchlings are extremely active at this time and, as they have a limited amount of available energy in their yolk sacs to get them well beyond the predator-infested littoral zone, we wanted to release them with minimal delay.

After 1971/2 we continued notching, selecting a different specific scute to excise each season, and by 1975 we had notched and released some 22,383 hatchlings. We were both optimistic and pleased, if not complacent, with the success of the programme. We were pleased that after many releases no hatchling with an excised scute was found dead along the beaches of Maputaland and we had no reason to believe that the programme was causing excessive mortality in the hatchling cohorts. What is more, we found hatchlings hundreds of kilometres south of the breeding beaches, alive and well, with the notch cleanly healed. The process was thus meeting our expectations.

As in previous seasons, we continued an extensive advertising campaign calling for volunteers to look for notched hatchlings all the way down to Cape Town. Both the SABC and local newspapers had been extremely cooperative. Some, like the *Eastern Province Herald*, went way beyond the enthusiasm expected with David Bickell, one of the paper's top journalists, becoming a turtle volunteer himself. I hardly need to add that the programme has received nothing but outstanding support and cooperation from all the aquaria along the South African coast: Durban, East London, Port Elizabeth and Cape Town, including the Two Oceans Aquarium, which was built many years later. The South African Museum was also an enthusiastic participant, especially Mr Reinhold Rau and his team of technicians, one of whom prepared the excellent leatherback exhibit still on view in the museum today.

In 1972, at the end of the hatchling season, I carried out an expedition down the entire east and south coasts of South Africa to search for hatchlings myself. En route to my selected main site, Cape Agulhas, I called into the Bayworld Aquarium in Port Elizabeth to see if they had found any hatchlings. As expected, Bayworld had a tank full of loggerhead hatchlings, which had been brought in by volunteers, and I quickly made my way, with a local colleague, to the tank where they were on public display. I asked whether any of them had been notched and was told, quite emphatically, no. Not a little disappointed, I kept scrutinising the little loggerheads when suddenly one banked sharply and there, against all odds, to my almost hysterical delight, was a notch. It was absolutely clear, the scar was totally healed and the hatchling itself appearing to be in perfect health.

This little animal, only 55 millimetres long, had made its way from Maputaland to Port Elizabeth, a distance of 1,100 kilometres, facing goodness knows how many dangers, probably barrelling along in the racing Agulhas Current before being blown ashore by an inclement wind, all in a matter of weeks. This was the first marked and released loggerhead hatchling in the world to have been found, and it confirmed that loggerhead hatchlings in southern Africa are dependent on the southward flowing Agulhas Current for distribution.

At times like this, it is perhaps worth reflecting on what is required from a research worker to establish a single fact worth having: thousands of hours of walking, experimental failures, bouts of self-doubt and disappointment, frustrated arguments with those who said a method would fail and then the long wait before the glorious moment that makes it all worthwhile.

## The Search for the Lost Decade – Part 2

This was one of those moments!

Over the years, many more were found, which proved beyond doubt that the Agulhas Current plays an essential role in the transport, distribution and sustenance of hatchlings (both loggerhead and, by association, leatherback) in the early months of their lives. The shortest time recorded between the release and recovery of a notched loggerhead hatchling was 18 days from Maputaland to Cape Agulhas, some 1,600 kilometres. Even more excitingly, the notching programme showed that there is a steady leak of loggerhead and leatherback genes around the Cape of Good Hope into the Atlantic Ocean. Notched loggerhead, and associated un-notched leatherback hatchlings, have been found in False Bay, Table Bay, Dassen Island and as far up the west coast as Elands Bay. Elands Bay is some 2,200 kilometres from the beaches where the hatchlings were released – an amazing voyage for a tiny animal that has reached only 65 millimetres in length.

However, I am getting ahead of myself. That single recovery, as well as the finding of many stranded hatchlings on the Cape Agulhas beaches during that trip, more or less convinced me that the whole notching programme was justified and effective. There was a whiff of complacency in the air as Colin Sapsford and I hunted turtles some four years later in Maputaland and came up behind a nesting loggerhead female with what appeared to be a notch visible at the edge of her carapace.

Colin's excited and somewhat imprecise question, which gives this chapter its subtitle, cut through the air like a knife as I stared at the 'notch'. I was suddenly struck rigid because there was no possible way that a loggerhead female could reach nesting maturity in four years. My Australian colleagues were postulating that loggerheads would not mature for 50 years or more. Even if they were wrong (which they were, but not by much, because their loggerheads have a much longer distance to travel than ours), this notched female turtle was a clear impossibility. It became obvious that the one-notch marking system had a serious flaw: single notches could be caused by accident. Colin, sensitive to my pain on discovering the technical flaw and eager to rub salt into the wound, said 'This bloody turtle can't be four years old!'

and, with some bitterness, I had to agree.

This discovery simmered for a couple of years as we contemplated the necessity of using more than one notch. In 1978 it was finally decided that the notching schedules for the coming hatchling season would be modified; henceforth every hatchling was to be notched twice in a combination of marginal scutes. The double notches provided a code that gave the total notching programme an experimental lifespan of 31 years before we ran out of unique combinations without excising more than two notches. The programme of notching was finally brought to a close in 2004.

All told, 347,000 loggerhead hatchlings were notched and released. Recent analyses by Jenny Tucek, one of Dr Ronel Nel's students, have demonstrated, from the many dozens of loggerhead females notched as hatchlings and recovered on the beaches as nesting adults many years later, that the Maputaland loggerhead turtles have an average age of 24 years at nesting maturity. After such a marathon effort, Jenny has provided another one of those moments.

Having demonstrated the role of the Agulhas Current in the movement of turtle hatchlings, it follows that wherever the currents lead, the turtles go along with them. Huge gyres are spun off the Agulhas and these find their way into the main gyre of water circulating in the southern Indian Ocean. This mass of water slowly spins, taking water eastwards at the southern edge, along the Antarctic Convergence and then up past the west coast of Australia and back towards Africa in the South Equatorial Current. The gyre carries, on its revolution past Reunion Island, the now sub-adult loggerheads, which, in parallel with the growth of the Maputaland loggerhead nesting population, have become the species and size most commonly encountered by fishermen of the island.

The total time elapsing between hatching and returning to the mainland of Africa is probably in the region of 7–10 years, with sub-adult turtles going through even more adventures before finally returning to breed along the beaches of south-eastern Africa. This is a truly amazing voyage, the discovery of which, today, could be described more easily and more rapidly, using the study of genetics.

This use of DNA to clarify relationships was very much in its infancy back in the 1960s and the fact that these amazing voyages were described using non-genetic tools is a credit to South African scientists. It must be admitted, however, that as thrilling as the near decade-long pelagic wanderings of the Maputaland loggerheads is, the voyage of our loggerheads pales in comparison with the truly gargantuan efforts of the loggerhead hatchlings of Japan. Although of the same size as our loggerhead hatchlings, they travel over 13,000 kilometres to Mexico and, over a similar distance, return back across the Pacific Ocean to sweep up to Japan via Okinawa and the Ryukyu Island chain – an awe-inspiring round trip covering more than 25,000 kilometres.

American colleagues, in cooperation with European scientists working in the Canary and Azores islands, have demonstrated a similar odyssey for loggerhead hatchlings emerging on the beaches of the eastern United States. They are carried towards Europe by the Gulf Stream and swept down the Atlantic past the Azores and the Canary islands, eventually making their way back via the North Equatorial Current to the Americas.

The more recent tracking of loggerhead hatchlings by fellow scientists in other parts of the world has been exciting, if not amazing. But, here in Maputaland, we successfully launched the largest and most ambitious hatchling-marking scheme in the history of turtle research and made a singular contribution to solving Archie Carr's mystery question. If it does nothing else, the study illustrates the instinctive courage and temerity with which small turtles enter the massive gyres of the world's oceans to spend perhaps decades trying to survive and reach nesting maturity. It must be admitted that sea turtles, over millions of years, have proved that their technique works.

This research result requires the thankful acknowledgement of the dozens of enthusiasts who walked the beaches of South Africa and collected hatchlings, notched and otherwise, fresh and sometimes singularly rotten, and patiently and, on occasions, under sufferance, despatched turtles to me. Although it is not possible to thank them all, there are three worthy of special mention: Peter Dreyer and Jock Dichmont of Cape Town (later of Arniston, or Waenhuiskrans as Jock preferred it called) and Piet van As. In addition to being a former engineer on the SAS *Agulhas*, Piet is a talented and enthusiastic artist. He almost always decorated his letters and envelopes to me with delightful cartoons showing either himself at his labours trying to find turtles or showing some disaster or other activity involving his ship.

With Piet's permission I have been afforded the rare privilege of sharing the fruits of his talent with readers of this book and have used a couple of them to add aesthetic value to the story of the search for the missing decade.

CHAPTER 9

# Amanzimbomvu to the Rescue and Other Tales

During the past 50 years or so, the turtle survey has gone from strength to strength with improvements in techniques, equipment and science, but, in truth, the role of the scientists could not have been fulfilled without the enthusiastic and willing support of so many non-scientists. These are the field officers, or game rangers as they were known in the early days, the game guards, the labourers, the mechanics, the friends and, of course, the local amaThonga.

In the beginning, one had to have a permit from the Department of Bantu Administration merely to get to the beaches, because Maputaland fell under the jurisdiction of that august body, magistrate and all. However, as the department had alerted the Natal Parks Board to the killing of turtles, they were sort of over a barrel. So the Natal Parks Board had to appear willing, over the first 20 years or so (before the department abandoned permits as a lost cause), to supply an unending list of requests for permits to cover students, staff and foreigners. Goodness knows how many thousands of permits lie rotting in some cellar, a mouldering testimony to the lunacy of bureaucratic control. One cannot imagine what the original objectives of such control must have been.

Heaven knows, there were enough other difficulties facing those intrepid enough to try visiting Maputaland. The first challenge was the roads – and here I am referring to the so-called main roads. In

1963 there were two routes to Kwangwanase: one from Josini via the Ndumu–Ingwavuma road and across the Pongola River at Makane's Drift; the other from Mkuze via Ubombo and across the Makatini Flats. Both routes were uncomfortable when dry and horrendous when wet.

The Ndumu route was often cut off when the Pongola River rose beyond the ability of the master of the pont, a local Thonga named Amanzimbomvu, to pole one and one's vehicle across the river. This did not always deter Amanzimbomvu from trying, as he was most often found in an alcoholic haze having wined since breakfast (or before) on the product of the local Lala palm. His name, literally translated, meant 'red water', which alluded to his constant suffering from malaria. His almost-frail frame was probably due to the same cause, but despite these minor shortcomings he was a local legend, available from dawn to dusk and always the source of a good grumble.

Just how dedicated he was to the task did not come home to me until New Year's Eve 1966 when one of our visitors to Bhanga Nek suddenly collapsed on the beach, squirming in agony – a few of his kidney stones had decided to go walkabout. Bhanga Nek was nearly 20 kilometres from Kwangwanase over the most appalling road. Rigging up a bed for the victim in the back of the old Land Rover was a challenge that was made worse by the fact that we had no canopy. Pat Temple decided that we should call on the Rutherford family for help. Bob and Dot Rutherford managed the Ndumu Limited store in Kwangwanase and were the parents-in-exile to all the students who worked the turtle beaches for the first ten years of the survey.

Shortly after midnight, with our visitor in excruciating agony, but warmly wrapped in blankets on the bed of the open Land Rover (thank goodness it did not rain that night), we set off for the store. There was no way that one could provide a smooth ride, but we did our best by driving very slowly and it took about two hours to get to the Rutherfords, who had a Land Rover Defender. Bob and Dot showed nothing but concern as we knocked them out of the bed they had not long occupied, due to their having enjoyed a New Year's Eve party of note.

## Amanzimbomvu to the Rescue and Other Tales

At about 3 a.m., having been seen by the local doctor, who insisted that the patient be taken to Ingwavuma, we set off with Bob at the wheel of his Land Rover. The patient was now in a sort of bed and at least out of the wind. In retrospect, I thought he was remarkably well behaved, just giving the odd cry as we drove over a particularly uneven patch or hit a pothole. When we got to the Pongola River, of course, Amanzimbomvu was nowhere to be seen. Bob knew where he lived, however, and the two of us plunged into the bush to find him. Personally, considering the old regime in South Africa, I did not expect our banging on his door at 4.30 a.m. to be greeted with cries of unmitigated joy and steeled myself to help Bob give him some ringing advice on duty, necessity and humanitarianism.

The door was answered almost instantly by Amanzimbomvu himself, who was, judging from his breath, also a recent reveller at a New Year's Eve party. Bob explained the problem crisply and, without a second's hesitation, Amanzimbomvu urged us back to the river, following at speed with a son in a state similar to his own. The river was high, quite alarmingly high, and we thought that the two of them would simply say it couldn't be done. Amanzimbomvu observed the river for less than a minute before telling us to get the vehicle onto the pont. This was a tricky job at the best of times, but with his guidance Bob got the Land Rover aboard and Amazimbomvu and his son poled us out into the rising river.

As the pont heaved downstream, I thought we were going to end up in Lourenço Marques (now Maputo) with a seriously stressed patient, but I did not fully appreciate the skill and strength of the four skinny arms poling us across. After an initial 50-metre downstream ride, Amanzimbomvu, with some deft shoves, angled us across the river to the shelter of the far bank and, with considerable effort against the current, he and his son poled us back upriver to the road on the far side. It was pitch dark on a flooding river and, without a murmur, these courageous souls got us safely across.

It was nearly 6 a.m. when we got to Ingwavuma and the hospital, into which we rapidly deposited our guest, who was in acute agony. We were absolutely exhausted, but could not leave until the fate of our patient was decided. We sat in the Rover and I pointed out to a very long-suffering Bob Rutherford that they had lit a fire in the hospital, which was emitting whitish smoke, not unlike that on the Vatican roof when they are electing a new Pope. Each time there was a change in the colour of the smoke I would cry out that a decision had been made. Bob, as I recall, was too tired to find this amusing and I believe he thanked the gods when the doctor emerged and told us that the patient would be all right. He was later flown to Durban where he recovered well.

The visit to Ingwavuma Hospital was quite fortuitous for me because, while waiting, I was stricken by a startling attack of diarrhoea (Bob said I deserved it) and asked the doctor for some help. He grinned, saying that he had just the thing. The hospital had just the day before taken delivery of the first batch of Lomotil tablets ever to reach Maputaland, a wonderful product of which none of us had ever heard and directly attributable to the needs of the United States space programme. It worked brilliantly.

During the long drive back to Kwangwanase and the turtle beaches, we contemplated how fortunate we were to have Amanzimbomvu come to the rescue with such compassion. He saw us back across the river with another fine display of pontmanship. Alas, the concrete bridge across the Pongola was built shortly after that and the pont and Amanzimbomvu disappeared from the scene for ever. His services, which had apparently lasted for nearly 40 years, were simply no longer needed.

The combined support of the Rutherfords, the doctors and the local amaThonga was truly indicative of the tremendous friendship and mutual warmth that existed between all the different parties in Maputaland in the early 1960s, when the area was truly wild and the amaThonga were not yet politicised.

Politics was slow in arriving in Maputaland and the whole area had an olde worlde atmosphere that was noteworthy for its lack of malice or resentment. A good example involved the local fishermen. The Natal Parks Board staff members were often drawn into local disputes or issues and we, from the turtle beaches, would get drawn in along with them. The saga of the lake channels was a case in point.

Boy Hancock arrived at the beach one morning and, kicking me out of bed, said I had to attend a meeting. Half asleep, Boy, Garnet Jackson (the officer-in-charge of Bhanga Nek at the time) and I drove over to Lake Hlange, launched the Parks Board's boat and headed for the channel linking Lake Hlange and Lake Pungwini (in order from the top, the second and third lakes in the Kosi Bay system). It appeared that the local Thonga elders had a request to make of the government. The magistrate from Ingwavuma had organised the meeting and asked the Natal Parks Board to bring along a scientist, if possible. As I was the most mature student and was well on my way to obtaining my BSc in zoology, Boy felt that I would fit the bill.

The basic problem was that no fish were being caught in the fish traps in the upper lakes. Although I did not realise it at the time, very few fish or prawns or anything else were being caught and productivity all over the southern Indian Ocean was poor that year. The primary cause was probably a La Niña event and even our sea-turtle numbers were low. Colin Limpus, in Australia, was in the throes of demonstrating that events such as La Niña and El Niño had a marked effect on the success of sea-turtle breeding seasons.

Such reasoning would have had little effect on the local fishermen, however, because they were convinced that the problem lay in the long and very convoluted channel linking the two lakes. In their view, the convolutions had become so complex that the fish could no longer navigate the channel. They wanted the government to dredge a straight and much shorter channel between the lakes, thus bypassing the existing channel.

When our boatload of delegates arrived at the gathering I was struck by the *Illustrated London News* characteristics of the gathering. (In those days the magistrate's nickname was 'The Lord of Ingwavuma'. Although this particular magistrate was quite benign, his predecessor, Mr Zeelstra, had earned the sobriquet.) The magistrate, very formally dressed and with a pith helmet, was sitting in a chair perched unevenly on one of the banks on the side of the channel, surrounded by his respectful staff and interpreters, while scattered around were about 20 fishermen. These were all elders and they stood there, tall and slim to a man, one leg wrapped around their 3-metre fishing spears without which, in those days, no Thonga fisherman was ever seen. No one was talking as our boat glided to the bank and the whole gathering reminded me of a Victorian drawing of Stanley meeting Livingstone on the shores of Lake Tanganyika.

On our arrival, the magistrate announced, with some officiousness, that the meeting had begun and the eldest fisherman, clearly a person of some influence, led their interpretation of events and made the request for government help. Shall I say that the proposed channel was not warmly received by the magistrate or indeed ourselves? In our opinion, whatever was causing a lack of fish, dredging a channel through the swamps was not going to solve the problem.

The magistrate shrewdly suggested that the Natal Parks Board staff should answer the fishermen. Boy Hancock, never one to dally, plunged into the argument with gusto. He asked the Thonga elder what on earth had given them this idea, as the channel had functioned perfectly well in its present disposition for years.

The elder replied with some dignity, addressing Boy by his local Zulu name of Izikhindi (meaning 'shorts', as Boy never wore anything else, and on formal occasions wore longs only under sufferance), stating that 'Before the war the channel was straight!'

Boy responded instantly and suggested, patronisingly, that the elder must be mistaken because he, Boy, and his father had visited the lakes to fish back in 1938 and the channel had been almost exactly the same as it was today.

The elder coolly replied, 'Not Hitler's war, Izikhindi! The Kaiser's war!'

With that reply, Boy was not the only one to be stunned. The elder statesman had put us all in our place. However, you could not put Boy down that easily. He promptly switched tack and said, '*Baba, Mnumsane!* If the channel has been modified like this over so long a time, this is clearly an act of God and the *Nkulunkulu* always acts in the best interests of the people!'

To which the Elder, without a smile, replied, '*Aykona*, Izikhindi. If the *Nkulunkulu* always acted in the best interest of the people He would *never* have sent us the white man!'

After about half an hour the meeting ended inconclusively, although I thought the magistrate, clearly lacking a sense of humour, looked a bit tense. Boy, Garnet and I were in hysterics in the boat on our way back to Bhanga Nek, allocating full marks to the wonderful old fisherman, who was a truly gifted negotiator. The channel remains today, perhaps even more convoluted, but fortunately later seasons saw the fishing return to acceptable levels of success and the suggestion was never made again.

Whatever the faults of the students and the staff of the Natal Parks Board, whenever there was a serious problem, the local tribesmen would come to Bhanga Nek for help. One of the most frightening events involved the Portuguese who, in 1966, were fighting a vicious rebellion in Mozambique. It appeared that a local Thonga had picked something up on the beach and, not knowing exactly what it was, but seeing it was wrapped in plastic and tin-shaped, he struck it with a cane knife, no doubt expecting to open it. Regrettably, it was a hand grenade, which exploded and killed him instantly.

A companion, who was nearby, ran to Kwangwanase and reported the death to the police, who shortly afterwards arrived at Bhanga Nek and warned us of the possibility of finding more on the beach. The packaging showed that it was indeed a Portuguese grenade. It transpired that it was from a troopship from Beira, which was homeward bound for Portugal around the Cape. We later learned that the ship had been subject to an intense search before leaving Mozambique waters to ensure that no excess weaponry was finding its way back to Portugal and these unused grenades, souvenirs or not, had been jettisoned. Those still in their plastic bags floated ashore along the turtle beaches.

The next morning, we went down to the beach and found one grenade and, while speaking to the locals and warning them of the danger of handling them, heard that another four had washed ashore. None of the students were wildly enthusiastic about carrying the one we had and, when we turned at the end of the patrol, Major Pat Temple turned to me and said, 'Give me the damn thing!' and sat himself in the back of the Land Rover cradling the grenade on his stomach.

Pat had a reputation for courage, having been the chief commissioner of police in Swaziland before taking early retirement and joining the Natal Parks Board, and many tales were told of his bravery in the face of real danger. One of the most striking was when there was a vicious riot in the prison at Havelock in Swaziland. The prisoners had taken control of the entire prison, killed a few of the guards and held the rest of the staff hostage. It was a very ugly situation and police and army units surrounded the prison. Many officers were hell-bent on blasting it to the ground and killing as many of the rioters as possible, but Pat, who envisaged the possible carnage, insisted on being given a chance to solve the problem without bloodshed. Armed only with his swagger stick, he marched alone to the gates of the prison and banged on them. When the gate opened, he disappeared inside.

After ten very tense minutes he emerged from the prison, the gates left open behind him, walked calmly back to the surrounding force, dismissed the army and told his men to please go inside and restore order. Not a single person was hurt or threatened during the process. That was the calibre of this immensely responsible man now bouncing along in the back of our vehicle with a live grenade on his lap. When we asked him why he preferred to carry the grenade, he replied, rather flippantly, that he thought it preferable to be killed outright if it went off, rather than be wounded by a piece of shrapnel if it exploded further away.

We were very impressed and our awe of Pat grew perceptibly. It was only later that year, when Pat died of pancreatic cancer, that I put two and two together. We knew that he was not 100 per cent well, but had not known, until long after this event, that he had cancer. He, himself, was probably certain of this and took charge of the grenade because he did not want to put any of us at risk. This was an act so typical of this gentle man. His death was a great loss to us, a great loss to his family and a great loss to the district. Pat was 52 when he died.

CHAPTER 10

# 'There's a *dog* nesting on the beach?'

Over the many years of the survey, one became accustomed to delightful surprises. There was always something new happening and, of course, there were many creatures other than turtles on the beaches. Many of them were sessile, adding beauty to the reef scene, but constituting a constant threat. One student, Joe Venter, had good cause to remember an occasion when he, careless of the falling tide and misjudging its depth, body surfed with great skill into a bed of *Tetraclita* on the rocks at Bhanga Nek. *Tetraclita*, I should explain, is a particularly sharp acorn barnacle that grows along the intertidal edges of surface rocks along the coast. Joe emerged half-flayed from his chin to his toes, his chest and legs scored deeply and his arms and hands dripping blood from the deep gouges inflicted by the barnacles. Doreen Brent, the wife of the officer-in-charge, spent half an hour cleaning the wounds, painting on copious quantities of mercurochrome and comforting Joe, who was in mild shock and trembling like a leaf. When he unstiffened enough to emerge from the shack, he was greeted with hoots of laughter from his grossly unsympathetic colleagues, as he looked like an animated Picasso.

Other animals were more of a problem to the turtles than the staff, however.

In the early days, serious damage was done to clutches of eggs and hatchlings by packs of feral dogs. A considerable effort was made by

### 'There's a dog *nesting on the beach?*'

the Natal Parks Board staff to persuade the local amaThonga to keep their dogs under control during the night. This effort achieved a degree of cooperation, but it did not eliminate the presence of dogs, nor reduce dramatically the amount of damage they were doing. Stronger measures were obviously called for.

Having dealt with jackals in the Giants Castle Game Reserve, I was quite gung-ho to apply some of my experience to the local canine problem. Jackal hunting with a pack of Peel hounds, as is still common in England, had been a very successful way of keeping jackals under control (if not just driving them out of the park) and I loved the experience. It was unthinkable, however, as it would have necessitated horses on the beach (see Chapter 4). Poisoning with 1020, a toxic substance widely used by farmers to combat predators, was not acceptable because it is non-selective and would probably have killed many innocent animals. Coyote-getters seemed to be the obvious answer.

The coyote-getter, developed in the United States, is effectively a short pipe containing a spring-loaded device for discharging a shotgun-sized charge of cyanide. The device is hammered into the ground and primed with a foul-smelling bait, which is attached to the trigger mechanism. When the coyote, jackal or feral dog bites and pulls the cotton wool, it discharges the cyanide into its mouth. It is very effective.

This method had been a great success in the United States. Happily, we had a new Natal Parks Board staff member helping on the turtle survey one year, David Rowe-Rowe, who was destined to become a mammalogist of note and later proved to be one of the most successful users of the coyote-getter. David was keen to help, but once again caution ruled the day and the Natal Parks Board refused permission to put out coyote-getters because of the possibility of local children harming themselves through curiosity.

So, it fell to old-fashioned shooting and for several years every vehicle patrol carried a rifle. As some of the feral packs consisted of up to 50 dogs, what sounded like a minor war exploded now and then on the otherwise quiet beaches. Success was limited, however, because shooting at night, with staff waving powerful torches in several directions at once, coupled with a horde of dogs scattering everywhere, was not conducive to the 'aim calmly and gently squeeze the trigger' school of hunting. Many shots were discharged, but not many dogs were shot. The general mayhem, along with one or two actual victims, however, appeared to do the trick and after three or four years the dogs mysteriously disappeared and ceased to be a problem.

So, when a beach guard arrived at Bhanga Nek one morning, looking quite alarmed, and reported that a huge dog was nesting on the beach about 30 kilometres south, we were a trifle surprised. That evening I accompanied Herman Bentley, the Natal Parks Board officer, on an intrepid hunt to remove the huge dog. We encountered most of the teams of beach guards on their foot patrols en route and the southernmost teams were flatly refusing to walk near the area where the dog had been seen. Near Island Rock, we were shown some tracks that reputedly belonged to the dog and they were big. The tracks even went from the sea to the dunes, but there was no sign of an attempt to nest. On closer inspection, although there was a superficial resemblance to a loggerhead track, I was certain that these tracks did not belong to a sea turtle. Drawing a sketch of a seal, I showed it to the guards, who agreed that the dog did look a little like that. Certain of our superior intellect and having a good laugh, we set off home. Cape fur seals had been recorded on the beaches in the past, so this was a matter of no great import.

It was an uneventful drive until we encountered two terrified staff near Dog Point. They refused to patrol further, saying they had been attacked just a kilometre or so further north. It was the first time that these two staff had encountered the dog and they were genuinely petrified. Herman and I loaded up the staff and drove quietly northwards until we found the tracks. They looked even bigger than the tracks further south and Herman and I followed them up the beach with care. When we got to the cliff of dune rock, there was indeed a large smooth lump on the sand. As we approached, the animal raised its head, opened its mouth and gave a modest grumble. It was a fully grown elephant seal.

What had happened to the two Thonga staff then emerged more

clearly. Guards patrolled in pairs, but tended to drift apart as one usually walked slightly faster than the other. The leading guard had come across the tracks and, thinking they were those of a loggerhead, followed them up the beach and encountered the large smooth lump lying on the sand. He had no idea what it was and switched on his torch to have a closer look. Unfortunately, his torch batteries were on the point of exhaustion, so all he achieved was a dull yellow glow.

While he was peering at the creature, his colleague arrived and switched on his torch, which had fresh batteries. The sound woke the elephant seal, which rose to a height exceeding 2 metres, opened its mouth, the inside of which is bright red, and gave an almighty roar as the bright torchlight hit it in the face. The one guard dropped everything and fled down the beach, taking his torch with him. The other guard leapt onto the cliff and scaled it as fast as he could, paying no attention to the razor-sharp projections commonly found on weathered dune rock. His hands later required treatment. We could see the two large star-shaped indentations on the sandy beach where his feet had been. Both Herman and I agreed that he had probably set the world record for jumping from a standing position.

The seal turned out to be a curious and harmless visitor for several weeks and eventually came north to Bhanga Nek. During the day, it was quite tame and allowed us to approach quite close before grumbling. It was a beautiful animal, clearly in good condition, but somewhat of a curiosity to the local amaThonga. One day, after having some sticks thrown at it by local children, it took to the sea and disappeared. The nearest elephant-seal breeding colonies are way south on the Prince Edward Islands, at least 2,000 kilometres away, so its visit had been a rare privilege. The guards boasted of the encounter for years and, alas, we have never seen another one in Maputaland.

Over the years, we have seen many different animals on the beaches. Snakes were very common, perhaps too common, in the old Bhanga Nek shack and I caught quite a few weaving their way through the thatch. A green mamba was in the roof of the bedroom one day. I was trying to manoeuvre it out of the thatch when I noticed that six people had come to watch the proceedings, which were, to me, rather tense. Mildly peeved, I referred them to the fact that there was only one exit, the door, and if the snake decided to make a break for it, it was likely to seek the same exit as themselves. The room emptied instantly.

During the day, we occasionally found snakes caught by the sun in the wash zone of the beach, trapped there until the upper beach sand cooled down during the evening. One puff adder moved minimally during an entire day, gradually following the cooling sand upwards as the tide came in. A 3-metre black mamba, caught in the same situation, reacted badly to our approach and started to behave very aggressively. Willingly placing discretion before valour, we took to the hot sand at speed.

We have seen hippos and crocodiles in the surf, and on the beach we have encountered water and land monitors, large-spotted genets, water mongooses, side-striped jackals, duikers and bushbuck, ratels and, of course, the odd beached whale and dolphin. Bottlenose dolphins passing by were always a pleasure to watch as they surfed in the incoming waves and the size of the passing schools of Cape dolphins was unbelievable. It is difficult to imagine a school of dolphins so large that it stretches from horizon to horizon in every direction away from the beach. But for sheer volume, the most unusual experience was the shrimp invasion of 1972.

Late one afternoon someone noticed that there were small shrimps, less than 5 centimetres in length, coming ashore. We went to look, picked up a few and threw them back into the sea, trying to attract the Arctic terns that were gathering in greater numbers than usual. As we went inside for supper, not one of us commented on the shrimps. Even after we had completed our evening walks and returned to Bhanga Nek, we noticed nothing unusual. Low tide was late that night so Barry Brent, the officer-in-charge, took the beach buggy out towards midnight when all the students were already home. As usual, he loaded up his game guards and left.

He returned within a short time, breathless with excitement, dragged

> *'There's a dog nesting on the beach?'*

us out of bed and cried 'Come and see this!' We all piled on the buggy and shot off down the beach. Within minutes we were driving in a veritable fireworks display of blue flashing lights that were emanating from the millions of shrimps being crushed under the tyres of the buggy. We took turns and drove to and fro. The spectators standing on the beach watched the entire buggy lit up in a ghostly blue phosphorescence and every tyre became a spectacular blue Catherine wheel spraying blue flashes several metres into the air. We were quite used to seeing, on relatively calm nights, the blue-green bursts of light in the cascading waves caused by *Noctiluca*, the plankton that is a common source of phosphorescence in the sea. These shrimps, however, gave us a display of bioluminescence on a scale unimagined.

For well over 30 kilometres down the beach, our drive was a spectacle out of a Disney film. Every passenger glowed blue as the Catherine-wheel tyres threw up shrimp fragments and coated us all with blue lights. When we passed the end of the shrimp stranding, it was as if the lights of the world had been switched off and even the buggy lights seemed ashamed to try to match the splendour of the light in which we had been bathed. It was without doubt one of the most beautiful and moving wildlife experiences I have ever had.

There were so many shrimps along that part of the Maputaland coast that the turtles did not come through to nest that night. They were either deterred from swimming through the stranding mass or had taken the night off to eat shrimp. The following morning, the beach, as far as the high-tide mark, was covered with shrimp up to half-a-metre deep. Every rock pool in the reef areas was almost solid with shrimp, with hardly any room for water. Thonga women and men flooded onto the beach with every receptacle they could find, scooping up shrimps by the tens of thousands. Many people grabbed handfuls of shrimp and ate them raw with obvious pleasure. It was a veritable shrimp orgy.

The common ghost crab, always visible on our beaches, appeared to treble in numbers, emerging from every nook and cranny of the beach to join the feast. Terns and gulls gorged themselves to a point where, clearly sated, they just sat down on the beach, showing a deep reluctance to fly even when one walked up to them. The sheer volume of the shrimp stranding overwhelmed everything and coloured the beaches and reefs a beautiful pink for over 30 kilometres.

This was nature at its most bountiful and its like had never been seen in Maputaland nor, to my knowledge, was it ever seen again. Within two days, all traces of the shrimp had disappeared, the beaches were swept clean by the tides and undoubtedly by thousands of predators. Our daily schedules returned to normal, but not one of us present that night will ever forget experiencing a truly awesome spectacle of bioluminescence of a magnitude probably seldom seen by humankind.

What happened in the sea, where the larger part of the shrimp shoal must have been before it came ashore, and probably afterwards when the survivors moved on down the coast, is a matter for some speculation. It must have been one of the largest windfalls of food ever seen by thousands of fish and other predators along the Maputaland coast. The carnage must have been awe-inspiring. What the shoal expected to achieve, and probably did, was to reach its destination with a trail of satiated predators overwhelmed by the sheer volume of food. One can only wonder at the ability of such creatures to survive the onslaught and still maintain viable numbers for reproduction.

This was a truly stunning example of predator-swamping, paralleling the survival strategy used more modestly by the sea turtles of Maputaland and the world at large.

CHAPTER 11

# 'Sea turtles are not consummated in the market!'

By the end of 1966, the turtles had become an exciting and absorbing study, which I did not want spoiled by diverting too much energy into studying for my BSc. Happily, I had chosen geography to support my zoology major. In my totally immodest opinion, this was a brilliant decision. I was blessed in the geography department by having some outstanding lecturers, among whom were Drs John McDaniel and Peter Tyson. Peter later became deputy vice-chancellor and professor emeritus at Wits University. These gentlemen went out of their way to teach students realistic and required skills and also, perhaps most important of all, to think out of the box.

Geography is a phenomenal subject because it covers virtually every element required for human survival: food, landscape, sea, economics, climate and meteorology. Only today, nearly 50 years later, is the world becoming aware of humanity's dependence on these components, the importance of which have been long neglected.

When we were informed that, for our third-year studies, we had to research and submit a junior thesis, I was delighted. I had spent four years as a game ranger with the Natal Parks Board and the movement to explore the potential value of wild mammals was just beginning to gather momentum. The Natal Parks Board was rapidly developing a justifiably sound reputation for wildlife management and I was an enthusiastic supporter of the concept. This policy was the beginning of what is known today as 'sustainable use' and was founded on the belief that what was of value to humans would ultimately prove valuable for animals. It is a simple concept that must be managed skilfully. So, I suggested to the geography department that the subject of my junior thesis should be 'Marine Turtles: A Geographic Study'.

In the department, this suggestion did not receive universal acclaim, but Prof. McDaniel and Peter were instructed to discuss the matter with me. It would have been safer to suggest a study of 'The Distribution and Production of Chicken Farms in Natal' or 'The Outcrops of Dolerite Sills in the Midlands'. Nevertheless, both lecturers heard me out as I explained that sea turtles played an important role in economics around the globe: they were the source of some highly sought-after foods, factories had been built to process them, their import and export involved dozens of countries, they were a common source of nutrition for many coastal communities and, in association with man, they had played a historical role in the past. I quoted Prof. Archie Carr, then the world's greatest authority on sea turtles, who claimed that 'the green turtle is the buffalo of the sea and the world's most valuable reptile'. They were not only convinced by my arguments, but reacted with growing enthusiasm to the idea. Delighted as I was with their responses, I could not help feeling that they had become, perhaps, slightly jaded by reading dozens of 'standard fare' junior theses over the years. My thesis would at least provide a change of scenery for them. Upon their recommendation, I was given the go-ahead.

With Prof. McDaniel acting as my supervisor, I lifted my researches into sea turtles to new heights and, to obtain additional original information, I despatched letters and questionnaires to 80 countries around the world. The response was amazing. I received replies from no less than 36 countries, covering species present, exploitation, legislative protection, uses and prices for products and, in a few cases, expressions of concern about the future of the resource. When I shared this information with my supervisor, he became genuinely interested and was enormously helpful in the production of my final thesis. On completion, the thesis was accepted and supported my degree.

The study inspired me to become far more involved in sea-turtle

## 'Sea turtles are not consummated in the market!'

conservation and provided me with a near-global perspective on the value of the animals to coastal communities around the pantropical world. What was particularly striking was the effort that so many people had made to respond to my queries. Some letters were received in Spanish, many in French and, of course, English – but some in English that had moved some distance from the famous Dr Murray's dictionary and had taken a life of its own. My favourite response was a statement from Vietnam that 'sea turtles are not consummated in the market!'

The thesis also provided a small, but significant, 'Damascus road' event when I was able to consider, for the first time, the possibility of seeking higher qualifications. The thought had originally been lodged by Prof. Jan Heeg of the zoology department in Pietermaritzburg when he visited the turtle beaches during the 1966/7 season with students and staff from his old alma mater, Rhodes University. I had taken the group out on a 'turtle trot' and had waxed fulsomely on the nesting turtles. On the way back, Jan quietly walked up to me and said, 'You will consider doing honours with us next year, won't you?'

And so it was that I did honours in zoology and passed, with tremendous help from Jan Heeg, Prof. Gordon Maclean and Dr Eddie van Dyke. Gordon, a bird enthusiast who is now famous for his ornithological work, proved to be a breath of fresh air. Eddie was a man with a deep love of knowledge and concern for students that, alas, became obvious to many of us only long after we left university and matured enough to realise that his rather unique way of lecturing was intended to make us think. He was passionate about getting his students to reflect and to question. We were always asking Eddie questions that he preferred not to answer because he was inspiring, or forcing, us to think. Looking back from 50 years away, I feel that the University of Natal in that era was incredibly successful in launching a generation of biologists that have made their mark in South Africa and abroad.

Meanwhile, in the rest of the world, conservation as a cause was beginning to develop into an effective force. The IUCN formed the Marine Turtle Specialist Group (MTSG) and the papers we published and the support given so freely by Archie Carr resulted in my being invited, through the Natal Parks Board, to the first meeting of the group in Morges, Switzerland in March 1969. As mentioned previously, I had been recognised as a sea-turtle specialist, probably only because they were desperate to up the numbers from 11 to 12.

It was pretty shaky ground for me and I was mildly terrified, as the meeting was being led by some very well-known biologists. Archie Carr had a global reputation as did Dr Tom Harrison, who, apart from his numerous decorations from the Second World War, was a popular author and benefited from the glory reflected from his wife Barbara, an anthropologist famous for her orangutan studies. Dr Leo Brongersma, from Leyden in Holland, was recognised as an outstanding taxonomist, Dr John Hendrickson was the pioneer of turtle studies in Malayasia in the late 1940s, and I had just graduated. I think the term 'small fry' described my position rather well.

The meeting started badly. Archie had asked us to bring a small slide show covering our local study areas and, as the new kid on the block, I was asked to give the first show. To give one an idea of how narrow our collective knowledge was at the time, my first slide showed the beaches of Maputaland and the clear offshore water adjacent to the beach. At once Dr Joop Schulz from Surinam interjected saying, 'This cannot be right because sea turtles only nest in dirty water!' The rest of the room turned and looked at him in wonder because we all believed that turtles only nested in clean water. Offshore of the Guianas the water is never clean, so Joop was right for Surinam. Having caused a bit of a stir with my first slide, I wondered what would next derail my nervous line of delivery. It did not take long.

The previous year, some Natal Parks Board staff had found a turtle carapace on the beach and it made its way to me. It was a very ugly carapace and appeared to have a twisted spine. I did not know what to make of it and thought that it might have belonged to an olive Ridley turtle, a species hitherto never recorded in South African waters. With the misplaced eagerness of youth to describe something new, I consulted my colleagues and then sent a picture to Archie Carr, asking him to confirm my tentative identification. Imagine my delight when

## 'Sea turtles are not consummated in the market!'

he wrote back and confirmed that it was an olive Ridley. The first I had ever seen.

Mike Mentis and I published a note in *Copeia* announcing the new locality record and, not unnaturally, I included a slide of the carapace early on in my presentation. When it hit the screen, several voices immediately said, 'What is that?' When I replied that it was an olive Ridley carapace there was a chorus of, 'No, it isn't!' Everyone there who had worked with Ridley turtles refuted the description and eventually consensus was reached among the experts that the carapace was that of a deformed loggerhead. I could feel my career and pretensions slipping through a crack in the floor and blurted out that I would write to *Copeia* at once to nullify the first announcement. One cannot imagine how I felt. I was making an idiot of myself on a global scale – one of the risks of being an immature scientist.

Archie Carr suddenly stood up, however, and in his Florida drawl said, 'Now hold on there, George. If anyone should publish such a notice it should be me. I was the one you turned to for help and I screwed up in confirming your identification. However, do not rush into print, because the only people that care about the truth are all in this room and we know it. Just get back there in due course and find a proper Ridley!' With the agreement of everyone, I was let off the hook and I had a feeling that was not unlike a really warm ray of sunshine coming through the clouds on a rainy day. It produced in me a glowing sensation and I knew that I wanted to be like Archie and the others. They had given me a lesson in forbearance and common sense that I believe has remained with me since that day.

Archie, of course, was my hero until his death in 1987 and I still regard him as both a mentor and a friend. He was an external examiner of my PhD thesis and I know that he was a generous and sympathetic judge, being ever helpful to young people.

It was shortly after returning to South Africa, having just joined ORI, that a gentleman arrived with a turtle hatchling in hand. He had found it while out for a walk on Warner Beach with his mastiff. The dog had gone off gambolling on the beach and returned to him, grinning as mastiffs do, with something flapping in its mouth. On investigation, he pulled a living, unharmed hatchling out of the dog's jaws and decided to bring it to ORI for identification. Well, the 'expert' looked at it carefully and announced that it was a loggerhead and a real mess of one too. It had too many lateral scales, was on the small side and even the head was a bit misshapen. The gentleman went off a little disappointed, I think.

The following day a lady arrived, also from Warner Beach, with two more hatchlings she had picked up while walking on the beach. I was about to pronounce as before when I turned one over. There were three odd pores in the scutes of the infra-marginal bridge that links the carapace and plastron of the hatchling. I grabbed the other one and it was identical. With growing excitement, I rushed the lady through the aquarium to a tank in which I had placed yesterday's hatchling and, sure enough, this one also had the infra-marginal pores known to occur only in Ridley turtles. Far from being ugly loggerheads, they were olive Ridley hatchlings, and they fitted the identification key beautifully. These hatchlings had obviously come from a clutch of eggs laid by a wandering female olive Ridley, as no other nesting record for the species had ever been found in South Africa.

It was as if she had emerged in a miracle move designed to save my reputation.

This emerging clutch of hatchlings was yet another 'baby in the taxi' accident, which, of course, is not to be decried because it is through these accidents that new colonies of turtles must have been founded all around the world.

We have, since then, caught the occasional Ridley in the shark nets, but they are indeed rare visitors to South African waters, although quite common in the Indian Ocean from northern Mozambique and northern Madagascar to India and Bangladesh. One of the greatest nesting concentrations of the species in the world gathers every year off the coast of Orissa, India, in the Bay of Bengal. Up to 600,000 females emerge over three days and, in recent years, these mass nestings, or arribadas (Spanish for 'arrival'), have been the cause of serious clashes between conservationists and developers as the Orissa government has given the go-ahead to develop a deep-water harbour in close proximity to these important nesting beaches. We await news of the effect of the harbour with some trepidation.

Of course, my presence at ORI was also a rare and fortuitous event. My ambitions to execute a survey of sea-turtle populations in the south-western Indian Ocean had been sparked and encouraged by friends and colleagues. How the research was going to be paid for was the problem. The Natal Parks Board had only modest interest in marine matters and saw such research as falling way outside their zone of jurisdiction and, although keen to continue supporting the Maputaland programme, had no funding for external studies.

Once again the gods took pity on me and an opportunity to obtain financial support suddenly became available. Dr Anton Rupert, with the blessing of the World Wildlife Fund, founded the Southern Africa Wildlife Foundation (subsequently changed to Southern Africa Nature Foundation, because wildlife did not translate easily into Afrikaans, and today is the WWF-SA) on 16 June 1968. This was a signal event and I immediately applied for funds and was fortunate in securing one of their first grants, which, alas, in May 1969, turned out to be R2,500, almost half of what I had asked for. The board had also taken the decision that the foundation would not fund the purchase of large capital items such as vehicles. I was suddenly faced with having some money to live on, but none to travel, which rather cut down the chances of executing the research programme. In addition, they insisted that the grant be handled only by a responsible institute and would not be available to individuals.

The Natal Parks Board had already declined any role involving research outside South Africa, so there I was, jobless, grant-less and homeless. The gods would have to find me a white knight, and they did, several of them in fact. The gods, however, renowned throughout history for their wicked sense of humour, also sent me a fairy godfather.

CHAPTER 12

# ORI and the Fairy Godfather

Towards the end of 1968, my girlfriend (or at least I hoped she was) asked me to help her assemble a display for ORI at an exhibition on the Durban campus of the University of Natal. She had graduated with honours in zoology in 1967 and regarded herself as incredibly fortunate to have a job at ORI. Any excuse to see her was taken with alacrity and I presented myself at the appointed time, received a string of instructions, and started working on the display. Today, I recognise it as a displacement activity that took my mind off the fact that I would be unemployed at the end of the next turtle season.

After an hour or so, she left me to get on with it and disappeared. In the midst of my labour, a man came up to me and asked bluntly, 'Who are you?' Somewhat taken aback, I explained that I was helping a young lady, whom I named. Having so clearly established my bona fides, he enquired about my position in life and my interests. He seemed quite sincere, so I blurted out all my current frustrations, along with a brief résumé of my turtle activities. He looked thoughtful as the light of my life reappeared, whereupon his attention turned to her and I was forgotten (she was, after all, a beautiful and intelligent young woman). I felt a pang of jealousy, but then I was introduced to the man, who turned out to be Dr Allan Heydorn, the newly appointed director of ORI. After a few more pleasantries, this very energetic and cheerful soul departed.

The following day, I received a telephone call to say that Allan would like me to see him at the aquarium as soon as possible. The meeting in his office was, for me, a very exciting one. He questioned me at great length about my work and aspirations and then said that ORI would, in all likelihood, be happy to act as administrator of my funds from the South African Wildlife Foundation, would be delighted to supervise my master's studies and would find me somewhere to work in the institute. Furthermore, he committed himself to trying to persuade the foundation to change their rumoured policy on funding to include vehicles. For the time being then, I had a home and the promise of some money, but if the rumour proved correct, nothing like what was needed to execute the proposed programme of research. I comforted myself saying *'toujours gai'*; half a loaf of bread is better than none.

Some months later, after the 1968/9 season, having attended the first meeting of the MTSG, I joined ORI. Allan welcomed me himself and allocated me a small space against a wall in the library with a half-moon table. For the next month or so, I sat in the library ordering scientific papers, reading such papers, studying techniques and becoming gradually more desperate as the foundation continued to dither over the vehicle decision. One day, sitting alone in the library, I was on the point of despair when my fairy godfather walked in.

He was a dapper little man in a pale blue open-neck shirt and blue shorts that were, in those days, not regarded as entirely decent. Sparkling clean he was, with a nice smile. He politely asked for the librarian by her Christian name. I explained that Blythe was on leave, but could I help? He told me she had promised him a book on piranhas and I tried to find one for him. I accepted his bona fides with the same sincerity that Allan had accepted mine, the common factor being that both of us had used a familiar staff name as an introduction.

Quite fortuitously, tea arrived and I invited him to have some with me. The conversation turned to me and what I did and I poured out my soul, ending only when my frustration about not having enough money for a vehicle was spat out. By this time I had tried many other sources to obtain a vehicle, with a spectacular lack of success, and had just that morning received another 'thank you for your interesting letter and request but . . .'

He listened with great sympathy and then asked, 'Do you mean that all you need is a Land Rover?' When I answered in the affirmative, he said, 'Heavens, I can do that!' A choir of angels burst into the 'Hallelujah' chorus. He explained that his aunt was the largest shareholder in the De Havilland aeroplane company and was always keen to make contributions to worthy causes when she visited new places. As luck would have it, she was coming out to South Africa in three months' time. He then proceeded to quiz me about ORI, its aims, financial situation and projects, ending the discussion with the statement that he was convinced that his aunt would almost certainly donate a substantial sum to ORI. With as much tact as I could muster, I praised his thinking, but reminded him of a more immediate and tragic situation. Ignoring my hint, he expressed a desire to meet some of the staff.

Within five minutes I had assembled three scientists and Allan Heydorn, to whom I had breathlessly and hastily explained the situation en route to the library. Allan swept in with the air of a Dale Carnegie – charming, welcoming and enthusiastic (that was genuine). Allan even ordered some more tea, a rare event at ORI. Within minutes, the fairy godfather had everyone enthralled as he explained about De Havilland and the fact that he was starring in the musical play *Robert and Elizabeth* (loosely based on *The Barretts of Wimpole Street*), which had just started its run at the Alhambra Theatre in Durban. We then took him for a tour of the aquarium and some of the offices and exhibits and an hour later I was once again alone with him in the library. ORI was abuzz with excitement.

So was the godfather, confiding in me that he thought one of my scientific colleagues a very attractive man, to which he added more specific positive attributes, the significance of which I generously overlooked at the time. I was, at this stage, fairly besotted myself, to which blissful state he added by phoning his local lawyer and telling him to set aside some money. In response to a question from the lawyer, he asked me how much a Land Rover would cost and I gave him a slightly inflated ballpark figure. He accepted it without any sign of distress and conveyed the sum to his lawyer, finalised his instructions and then asked me to join him for lunch.

After the lunch, for which he paid, he requested my telephone number and for directions to McCarthy's Garage, which was barely a kilometre away in Smith Street. He departed on foot to go and see about my Land Rover, spurning the offer of a lift, saying the walk would do him good. About an hour later, he phoned and said the only Land Rover of the model that I had described was baby blue. Would that be acceptable? Bubbling with excitement, I pointed out that colour was the very least of my concerns. He then asked if anything special should be done to the vehicle and I suggested that it be Corrosolved (sprayed with a rubber solution that would protect the metalwork from rust). He hung up after saying he would be back for tea.

On arrival, he had a brochure on the Land Rover and we sat together looking at all the finer details. It was marvellous and I noted that the salesman's name and telephone number were on the brochure. He told me that the vehicle had been sent off for its protective rubberising, but would be back at about 5 p.m. and he would go back and collect it for me. Once again I offered to take him, but he insisted on the walk, saying that he was enjoying this whole episode so much. He asked about the licensing of the vehicle and I said that it should be licensed in my name. He carefully took down all my details and said that it would cost R10, which, as it was to be my personal vehicle, I should pay. I gave him the R10 and off he went cheerfully, asking me to wait for him at ORI in case the vehicle took a little longer than expected. He left me in a joyful stew of anticipation.

Into my little bubble of euphoria stepped my good friend and ex-turtle colleague, John Bass, who was also at ORI reading for a PhD in zoology, his field of specialist interest being sharks. I think he had absorbed some of their more cynical characteristics because he announced that the godfather was a fraud. He was booed down by everyone, including me, but he remained resolute in his assessment and, may he never be forgiven, insisted on staying with me after the 5 p.m. deadline had passed and everyone else had gone home. The phone rang at 5.10 p.m. It was the godfather who said that there had been a delay and that he had an appointment to collect the Land Rover at 9 a.m. the next morning. John, to my rising ire, immediately said, 'I told you so, he is stalling!'

The godfather had given me his address and telephone number, but I suddenly realised that I had no name, and, what with John's discouraging, and truly hateful, refrain that he was a fraud, I foolishly told John where he was living. 'Let us go and see him!' he cried, saying that he would drive me home afterwards. We first tried the phone and received the dolorous voice of the telephone service saying that no such number was listed. That clinched it and the two of us went to the address given. We knocked on the door of a flat in Addington only to have it opened by on old lady who said that she lived there alone. My bubble of euphoria began to leak.

The next morning I was at work very early and, after seeing Allan, and carefully avoiding John, I phoned the garage and asked for the salesman. He immediately greeted me with a cheerful cry, 'You are one lucky guy.' He told me the vehicle was ready and waiting, everything done, with only the papers requiring a signature. He then confirmed that the godfather was arriving at 9 a.m. I gave him my number and asked him to phone me the moment he arrived because I had some doubts about him. 'Nonsense,' he cried, 'a nicer bloke with so much enthusiasm for your work I have yet to find!' My bubble stopped deflating, briefly. At 11 a.m. the salesman phoned and said, 'I am going to *kill* that bastard.' My bubble burst.

In a show of solidarity to a fellow idiot, I went up to McCarthy's and met the salesman and, of course, had a wistful look at the sparkling Land Rover. This sad tableau was not a period of calm as we jointly damned the godfather and wished him a swift despatch to hell. It was not a memorable conversation, but I did resolve to try to find the cause of our pain. The godfather had mentioned, in passing, that he was a regular patron of the pub at the Beach Hotel when he came to Durban. That night John and I staked out the pub in vain.

The following day, I extended my search to the Alhambra Theatre by phoning the manager and patiently explaining what had happened.

I described the godfather and asked if my description matched any of the actors in the play. To my amazement, instead of bursting out laughing, he was very sympathetic and invited me to attend a show, saying a guest ticket would be awaiting me when I chose to come. I said that I'd be there that very night. I duly arrived on time, collected my ticket, missed the manager and went in and enjoyed the show. I had taken a pair of binoculars and scrutinised every actor as he appeared on stage. To be honest, I had a good look at some of the actresses too. But, alas, no actor even resembled the godfather. On the way out I met the manager, thanked him profusely and explained that I really did not want to charge the guy, or even hit him, but merely ask him why he picked on a poverty-stricken biologist to practise his skills as a con man. The story did not end there.

Two days later the manager of the Alhambra phoned me to ask whether I had had any luck in tracing the godfather. I replied in the negative and thanked him for his thoughtfulness, especially given the fact that all I wanted to do was confront the godfather. There was a burst of rage from the manager who said, 'You can be as academic as you like, but if I catch him I will see him locked up for the next 20 years!' When he calmed down, he explained that, in his absence that morning, the Alhambra Theatre had been contacted by the Durban Technical College department of fine arts. When the manager returned the call to the professor of fine arts it appeared that, some weeks before, he had been visited by a dapper little dark-haired man from the theatre to borrow some of their museum's valuable period art works and sculptures for use on the stage settings of *Robert and Elizabeth*.

Neither the artworks nor the fairy godfather were ever seen again.

For the next few weeks at ORI, I bathed in the role of the victim, thus keeping my girlfriend's sympathetic ministrations close and personal. It took a few months before the foundation announced, in October, that it had changed its policy and that a further R5,500 had been allocated to my project. Many friends had sent letters to the foundation in support of the project: Dr Fritz Vollmar from WWF International, Dr Colin Holloway from the IUCN, Colonel Vincent from the Natal Parks Board and, of course, Allan from ORI, and I am convinced that the appeals from my white knights eventually won the day. I rushed out and bought a second-hand Land Rover panel van from one of my best friends, Pat Acutt. I was to spend a great deal of time with him, as he also kindly offered me the free use of the Wendy house in his garden. This was ideal for me as I intended to be away for long periods of time on field trips and a rented flat would have been money down the drain.

Within days, the Land Rover was at ORI, promptly christened the 'Panda Wagon' because of the large WWF panda stickers that I had to place on the doors, and my spirits were soaring. At last I had a home, a job and financial resources and could start planning my work at ORI.

CHAPTER 13

# How Hatchlings Live and Die and Other Dangers

Working at ORI was a real privilege and joy. My fellow scientists and the general staff were a very tight-knit and happy family and Allan Heydorn was a deeply respected and well-loved director, who was given to practical jokes. He was also a willing supervisor and was delighted when I announced that I wished to undertake experiments to try to establish how hatchlings survived in the cold waters of the Cape. This necessitated my booking the temperature-controlled tanks at ORI for quite some time.

In autumn 1970, I returned from Maputaland with a large box of loggerhead hatchlings and designed a programme to test their tolerance of, and reaction to, different water temperatures. The experiments involved keeping test batches of hatchlings in water of differing temperatures, checking their food intakes and monitoring their responses to temperature variations. Since 1967 I had received reports of hatchlings washing ashore along the beaches of the Cape, especially Cape Agulhas, and I wanted to know why they were doing so.

Loggerhead hatchlings are completely dependent on the warm Agulhas Current for their initial distribution along the South African coast. When they enter the sea in Maputaland, the water temperature is at its highest for the year, reaching nearly 26° Celsius. The Agulhas Current also reaches its maximum speed, between 5 and 9.3 kilometres per hour, at this time and flows relatively close to shore, as little as 3 kilometres from the beach. To further fulfil its role as the caretaker of a new generation of sea turtles, the current also carries its heaviest concentration of pleustonic, or upper-surface, animals such as the purple storm snail, *Janthina*, the common or garden blue-bottle, *Physalia*, and many other species of Cnidaria such as *Velella* sp. and *Porpita* sp.

With my first batch of hatchlings at ORI, I tested their response to colour using small painted Styrofoam blocks. Blue won at a trot and it was clear why. The hatchlings' primary food sources in the Agulhas Current were all various shades of blue. It is interesting that blue has evolved as the most useful colour to protect these species from over-exploitation, while parallel evolution has given the hatchlings a guide to these abundant food sources and the tools to deal with them.

Loggerhead hatchlings are equipped with two sharp claws on each fore-flipper. When a hatchling encounters any of their food items, it bites the animal and then digs these four claws into the prey. The hatchling then jerks its head backwards, using all its weight, and rips a piece out of its victim. I should add that loggerheads are totally impervious to the stings of bluebottles and other jellyfish, so the secondary protective mechanism of these species, the stinging nematocyst, is ineffective. One adult loggerhead was observed over the Agulhas Bank munching its way through a massive raft of bluebottles with equanimity. Such agglomerations of bluebottles are not uncommon, especially in the late summer when the individual bluebottles have attained their greatest size. The mere sight of one of these rafts is enough to give me cold shivers, having been stung many times by single bluebottles during my snorkelling activities, but loggerheads dine without pain as millions of nematocysts are discharged against them.

The same applies for the leatherback turtle, whose main diet consists almost entirely of members of the jellyfish family. In fact, leatherbacks are uniquely adapted to dealing with jellyfish. They have a lower jaw ending with a sharp point, almost a spine, which fits neatly between two sharply pointed downward-projecting bony extensions from the upper jaw. This three-pointed jaw structure enables the leatherback to puncture, grip and rip apart the normally slippery jellyfish. The leatherback has an oesophagus lined with inward-pointing cartilaginous spines, which presumably prevent the pieces of jelly-like food from being regurgitated when the turtle is feeding at depth. It is quite astonishing that this massive animal derives its sustenance from jellyfish, most species of which consist of about 98 per cent water. Probably as a result of the fact that its primary foods are nutritionally poor, leatherbacks swim incessantly and feed as long as food is available. Satellite-tracking of adults has indicated that they seldom stop and can quickly move considerable distances, probably searching for jellyfish shoals.

Unfortunately, leatherback hatchlings are notoriously difficult to rear and no one has been able to raise them beyond the age of three. They never adapt well to tanks, and special precautions, such as soft plastic barriers, must be built into the tanks. If this precaution is not taken, the hatchlings continuously swim against the walls until they have scraped their skin raw and eventually die. No attempt, therefore, was made to include them in the temperature experiments at ORI, despite the fact that they too wash ashore along the coast.

The series of experiments proved that loggerhead hatchlings can tolerate quite a wide range of temperatures, living comfortably and feeding with vigour at temperatures of 20°–26° Celsius. In this situation, which is similar to the environment in which they find themselves once they enter the Agulhas Current off the beaches of Maputaland, they grew well and put on weight rapidly. At this stage they are threatened by many predators, from cuttlefish to sharks, and there must be considerable evolutionary pressure for them to grow rapidly, as every few centimetres of growth reduces the number of predators able to eat them.

In among all the food items in the Agulhas Current, hatchlings are threatened by a small, but vicious, predator called *Glaucus atlanticus*. Barely 3 centimetres long, and also bright blue, it is a member of the snail family and swims by means of four extended fan-like fins. When it latches onto the flippers of a hatchling, normally the hind-flippers, it eats through the flipper exactly as a caterpillar does a leaf, carving off flesh with its radula and leaving clear half-moon-shaped wounds.

These wounds, although healed, remain through to adulthood and could be, and probably often are, mistaken for shark bites.

With a survival rate to adulthood in the region of two in 1,000, there are clearly many predators feeding off these 45-millimetre hatchlings. One giant kingfish, *Caranx ignobilis*, was caught off the Kosi Bay mouth with no fewer than 17 whole hatchlings in its stomach. To make matters worse, adult loggerheads have been known to cannibalise hatchlings. There is no room for sentiment in a loggerhead's world. Once the eggs have been laid, they are on their own.

One set of experiments showed that at lower temperatures, below 20° Celsius, the hatchlings became sluggish and seemed to find feeding difficult. When the temperature was dropped further, to 14° Celsius, the hatchlings stopped feeding entirely. They folded their flippers over their backs and simply floated. They lasted only two weeks at this temperature before dying. After the first few died, survivors were removed from the cold water and acclimatised gradually through a series of higher temperature tanks until they reached the ideal of 24° Celsius plus, where they re-established a vigorous feeding regime and started to grow again.

These results explained why so many hatchlings, up to 400 at a single stranding, could be washed ashore on the coast near Cape Agulhas. The cause is almost certainly an upwelling event when the powerful south-easterly winds hit the Cape region. The wind blows the warm Agulhas Current away from the coast and cold water rises in its place. This is not a smooth process, so patches of cold upwelling water will catch hatchlings as they are swept by on the inner side of the current. The temperature shock will simply stun them and weaker animals will almost certainly die. Others will become moribund and just float, being unable to feed. If the wind changes, such hatchlings, incapable of resisting, will be blown ashore.

After a real buster in 1971, hundreds of hatchlings of both species were washed ashore at Cape Agulhas. Many were already dead, but others had warmed up in the sun sufficiently to try to regain the water from the strand line, perhaps 50 metres up the beach. The sea was so cold that those that made it to the water immediately turned back onto the relatively warmer sand. By the time I found the stranding site, all had died, but the signs of their desperate wanderings were clearly etched in the sand. The gulls must have had a field day.

Studies have shown that these upwelling events are quite regular in the inshore waters of South Africa, but that the central core of cold water seldom endures for more than 9 days before local currents break it down and mix it with warm Agulhas water.

It is clear that our loggerhead hatchlings are genetically adapted to face these environmental threats. They are able to be caught in an upwelling event and endure the experience for up to 14 days, which should be more than long enough, provided they are not blown ashore. Once feeding again, despite the stoppage in growth for ten days or so, our chilled captive hatchlings, now markedly smaller than their more fortunate siblings, grew as rapidly as those that had never been exposed to the cold. The net result is that one can have hatchlings of the same age exhibiting a wide range of sizes and weights. A couple of weeks in close proximity to an upwelling event and a single cohort of sibling hatchlings will show a very wide range of sizes. This is clearly why the size of adult loggerheads is no indicator of age.

This is not the only problem that faces loggerhead hatchlings, however. All hatchlings are capable of diving only a short distance and because they float in a soup of other organisms, many of these organisms hitch a ride. By the time our Maputaland hatchlings approach the Cape, many are already weighed down by colonies of goose barnacles, acorn barnacles or sheets of Bryozoans. There is little doubt that the chances of survival of a heavily infested hatchling will be reduced.

Human waste, in the form of plastics, is another threat to the hatchlings as they will, by nature, bite anything that appears even vaguely edible. Many hatchlings were found to have eaten small plastic pellets, which seemed to have been dumped into the sea in their millions. There is no evidence to suggest that ingesting the pellets is fatal, as the hatchlings that I dissected were probably all alive when they stranded. But, if surviving hatchlings at sea continued to ingest them, the results might well prove undesirable.

Adult turtles face a serious threat from plastic bags or sheets of plastic, which can easily block the gut and cause the animal's death. In 1971 a leatherback turtle was caught in the shark nets, extricated and brought to ORI. It looked fine and was dropped into the main tank, but it survived only for 24 hours. On dissection, we found water in its lungs and it had clearly been on the point of drowning when brought ashore. Further dissection found that its large intestine was filled with a plastic sheet. When the plastic was cut out of the gut, it was obvious that it had been there for a long time. Peristalsis, the automatic movements of the alimentary canal that drive food through the system, had gradually compressed the plastic sheet into an almost-solid 2-metre roll of plastic some 5 centimetres in diameter. It took two of us, exerting real effort, to unfold the roll to its original size. It was a sheet of heavy plastic measuring 3 x 4 metres.

What is regrettable is that the spines in the oesophagus, designed to help the leatherback retain its food when feeding at depth, work against it. Once a portion of plastic, such as the one described above, has entered the spiny oesophagus it cannot be regurgitated and the leatherback can do nothing but swallow it completely.

The good news in this case, judging from the presence of food particles right through the system, was that the rapidly digested jellyfish particles were bypassing the plastic. There seemed to be little doubt, however, that a few more plastic sheets like this would have caused a full blockage. Smaller sea-turtle species are at even greater risk and plastics are probably taking a serious toll.

The ORI experiments on hatchlings also provided some interesting insights into the loggerheads' ability to grow. It is a popular belief, probably due to the publicity given to ancient tortoises, that turtles are very slow growing and that their growth is a steady process. This was found not to be the case. When hatchlings were moved into more spacious tanks, they demonstrated an ability to grow incredibly rapidly. Even though the food provided and temperature of the water were the same, they grew at almost twice the speed that they had been growing in the smaller tanks. This provides clues to the ability of hatchlings to grow rapidly in the Agulhas Current where there is plenty of food and plenty of space, with probably little competition between hatchlings.

As mentioned before, there are great survival advantages for hatchlings to grow rapidly from the diminutive soft-shelled stage, when they are 45–200 millimetres in length, to a size exceeding 200 millimetres. At this stage, the carapace begins to harden and the trailing edges of the scutes grow into hard, sharp spines that form a defensive barrier against most pelagic predators. The spines persist for probably six to eight years before the entire carapace widens and draws the spines back down to the flat carapace scutes common to adult loggerheads. By this stage, loggerheads are adept swimmers and probably capable of avoiding most predators.

As the survivors of the tank experiments grew up, we were eventually able to sex the loggerheads. As hatchlings, there is no way, short of sacrificing the animal, that sex can be externally identified. When a loggerhead reaches about 60 centimetres in length, however, the tail of the male begins to lengthen rapidly, eventually projecting well beyond the edge of the carapace. The tail of a female loggerhead, as with all other species of sea turtle, hardly projects beyond the edge of the carapace. It is, from then on, very easy to sex adult turtles.

One final lesson we learned was that one of the two claws on the foreflippers of hatchling loggerheads, so important for their early feeding, is lost at about the same time as the sexual characteristics become apparent. In adult females, the surviving claw has little function, (other than to seriously scrape the arm of a turtle-trotter from time to time), but in males this claw strengthens and develops a substantial hook shape, which is used to hold onto females during mating. Older female green turtles often have deep notches in the front of the carapace as a result of the attentions of males during mating.

Sex is a serious business for sea turtles and once *in copula* the process can last for four days. It is not a restful time, as unsuccessful males are constantly trying to dislodge the copulating male. The lucky male has his problems too because if the female decides to sink to the bottom, he just has to hold his breath until she is ready to surface again. The options open to him are clear: either risk drowning by holding onto the female or risk losing a hard-won opportunity to pass your genes on to the next generation. If he is fortunate, he will impregnate the female with sufficient sperm to last for several seasons, as females are capable of storing sperm and releasing them when required to fertilise a batch of ova. Recent research has demonstrated that this ability allows females to store sperm from different males and thus broadens the chances of successful egg fertilisation – yet another remarkable survival tool.

CHAPTER 14

# Guiding Turtles to and fro and Other Wonders

In the 1960s, one of the many challenges facing turtle biologists was the question of how sea turtles found their way from one end of the earth to the other. We had the proposed solution, of course: they used the sun, the moon and the stars. Archie Carr and one of his colleagues postulated that green turtles, moving some 1,400 kilometres to and fro across the Atlantic Ocean on their nesting migrations to Ascension Island, were possibly following olfactory cues. This was an interesting theory, but somewhat handicapped. Ascension sits in the South Equatorial Current flowing from Africa to Brazil. Most of the Ascension green turtles swim from Brazil to Ascension and back, so, if the turtles were following the scent of the island, it could not be followed in both directions. Archie, clearly aware of the theory's limitations, implied that the return of the animals to Brazilian waters after nesting might be a passive swim in the South Equatorial Current. Similarly, the possibility of sea turtles using celestial guides was not supported by the fact that sea turtles cannot see much detail out of water, as their eyes are adapted to underwater vision.

When we thought about how our loggerheads and leatherbacks were finding their way to the beaches of Maputaland, we stumbled over the same problem. We have recovered tags from loggerheads from as far north as Somalia, as far south as around Cape Agulhas and into the Atlantic and as far east as Madagascar. Leatherbacks have shown a similar, but much greater, range of distribution, as many end up in the mid-Atlantic. There is simply no way that any olfactory cue can be distributed in all directions and provide a series of markers that can be followed by a sea turtle.

The whole concept of genetic mapping (an animal's ability to know where to migrate without being taught) was in its infancy at the time and, having neither the skills nor the equipment to pursue studies in this field, we had to await the work of other scientists to try to solve the problem of long-distance migrations. Thanks to the excellent studies by Dr Ken Lohmann and his colleagues at the University of North Carolina, it would appear that sea turtles are capable of using lines of magnetic force, rather like a Global Positioning System (GPS), to identify where they are and where they are going. Quite how they do it is not yet fully understood, but the theory certainly meets all the requirements asked of it. It is multi-directional and needs no immediate physical association, such as chemical or olfactory cues, celestial markers or ocean currents. It appears to be the only sensible explanation for the sea turtle's ability to traverse thousands of kilometres, from any direction, even across oceanic boundaries, and still find its way to its natal beach as an adult.

This instinctive knowledge of a specific locality, a nesting beach in this case, is a remarkable phenomenon. My own theory is based on the fact that the turtle hatchling takes approximately two months to develop in the egg before hatching. During this period, the mechanism for setting the coordinates of the nesting beach, whatever it may be, fixes the global position in the body of the turtle. Like so many pieces of information hard-wired into a turtle hatchling, it is not a conscious talent that can be manipulated or read at will. If it has been properly hard-wired, the turtle will automatically find its way back to where it was born, just as it automatically knows how to dig a nest and lay eggs. What is even more surprising is that turtles, certainly loggerheads, appear to be able to fix the coordinates of their feeding grounds later in life. This is especially impressive as these feeding grounds may be thousands of kilometres from the beaches on which they first emerged.

There are some problems with the theory, however, in that the lines of magnetic force move as the magnetic poles shift. We do not yet know how the system works, or where the receptors activated by the force-lines are situated. They do not appear to be situated in the brain, however. Prof. Floriano Papi, his excellent colleague Dr Paolo Luschi and others from the University of Pisa, have fitted force-disrupting magnets onto the head of experimental turtles and many of them have found their way back to nesting beaches after being displaced for hundreds of kilometres.

In one elegant experiment, proposed by Floriano and Paolo, two loggerhead females were captured at the end of the season in Maputaland when they had completed their nesting. They were transported by road to Durban and, after a short period in captivity at ORI under the care of Rob Broker, a Natal Parks Board officer, they were loaded onto a freighter that set sail for Mauritius. Each was fitted with a satellite transponder en route and then released at sea. One was released immediately south of Madagascar, about 1,000 kilometres from the coast of Africa, and the other 100 kilometres or so from Reunion Island, some 2,000 kilometres from Africa. The results of this experiment were amazing.

The first female, released just south of Madagascar, spent about 24 hours in her site of release and then turned westwards, swimming an unerring straight-line course for the coast of Africa. Within 100 kilometres or so of the African coast, she turned due north and swam up the coast until she was offshore of Beira. There she settled into the pattern of behaviour associated with the return to a feeding territory.

The second female, 1,000 kilometres further away, with the land mass of Madagascar between her and Africa, appeared to be somewhat confused. She turned north and, well away from the coast of Madagascar, swam for a couple of weeks until she was two-thirds of the way up the body of this great island. She clearly began to suspect that all was not well and turned south again, swimming steadily, over 100 kilometres from the Malgache coast, until she reached the same spot where the first female had been released.

She then turned westwards and followed almost exactly the same route as the first female, heading for the coast of Africa. The really admirable moment came when, halfway across the Mozambique Channel and still several hundred kilometres from the African coast, she turned sharply north and swam a straight-line route up to northern Mozambique before approaching the coast. She then swam out to the edge of the Comoros and then back to the African coast near Tanzania and followed it all the way to Zanzibar, which was, from her subsequent behaviour, clearly her feeding territory.

This loggerhead female, at the end of a breeding season, had swum, virtually non-stop, for 5,000 kilometres. She had corrected a displacement of 2,000 kilometres and catered for an obstacle like Madagascar, of which she probably had no knowledge or experience. Now that, by any standard, is a remarkable achievement and demonstrates how powerful is their ability to navigate over long distances and unknown territory.

A year later, the Pisa team, assisted by staff of the Natal Parks Board and with the help of Unicorn Lines, displaced a group of three loggerheads at the end of the nesting season. The females were transported by road to Durban as before and held for a short time in the tanks at ORI. They were then loaded onto a freighter and all three were released slightly to the north of Zanzibar, about 200 kilometres from the mainland. This time only one swam south, with conviction, to the coastal waters of Zanzibar, where she clearly settled into her feeding territory. This meant that she had got home after the season with a great deal less effort than had she swum all the way from Maputaland.

The other two swam north into the northern Indian Ocean. Clearly lacking direction, both eventually turned towards the Kenyan coast and crossed the Equator. By then they were several hundred kilometres apart. Once over the Equator, however, they, quite independently, turned eastwards, swimming across the Arabian Sea to the Maldives. Their paths coincided, in due course, and, within eight days of one another, they swam a route between the same Maldives islands and headed out into the Bay of Bengal. The approach to the Maldives was almost a straight line – a mirror-image of the route taken by the females south of Madagascar the year before. Could it be that crossing the Equator reversed their ability to detect their position? Alas, we shall never know, as the batteries of both transponders failed shortly after entering the Bay of Bengal and the loggerheads swam into yet another turtle mystery.

Long before this, however, in Maputaland in 1970, we tried to ascertain how loyal turtles were to sections of the beach. Did they return to the same nesting site during each season and did they return to the same nesting site in other seasons? In retrospect, without knowledge of the exact mechanism of navigation, we were trying to see just how accurate and dependable their hard-wiring was. We decided to set up a series of permanent markers, 400 metres apart, along 56 kilometres of beach. We estimated the distances, either north or south of Bhanga Nek, where each turtle's nest was recorded. The marked poles, put in place through the hard work of Ranger Barry Brent, simplified the site recordings immensely and they remain in place to this day. Barry was one of the many Natal Parks Board officers who regarded the turtle programme as a holy grail and worked his heart out for it for four years.

As the seasons passed, we started to get a much better idea of the nesting distribution along our beaches and produced a much improved set of quantitative data, enabling us to track the nesting behaviour throughout each season. On average, loggerheads lay 116 eggs per clutch and nest four times per season. One of the first things that we learned was that they do not return to exactly the same nesting site each time they emerge, but tend to re-nest within a stretch of beach about 10 kilometres long. When one considers that a female may have swum 2,500 kilometres to Maputaland, this remains a remarkable feat of navigation and would certainly endorse any reasonable argument that they are hard-wired to find a fairly narrow stretch of beach. There were two other aspects of loggerhead nesting behaviour that the more detailed sets of data provided.

After compiling about ten years of nesting records, and plotting them against a schematic map of the 56 kilometres of beach, we noticed that there were two main sites of emergence, both adjacent to the Kosi lake system. The majority of the loggerhead nests were found in these sites. On closer inspection, we noticed that there was no normal distribution curve straddling each major site. Both had more loggerheads nesting to the north than to the south. The question, then, was why?

The east coast of South Africa is characterised by a north-moving inshore counter current, which is present, certainly along the Maputaland coast, on the shoreward side of the south-moving Agulhas Current. As both intense nesting nodes were clearly associated with the Kosi lake system, and we were aware of the underground freshwater streams moving from the lakes to the sea over the impermeable dune rock on which the beaches lie, could it not be possible that the water entering the sea was carrying a scent or olfactory cue? If there was a scent-recognition mechanism imprinted during the two-month gestation period, associating the scent with a safe nesting site, the identification of the scent would assist the turtle in 'recognising' a satisfactory nesting site.

If there were two significant sources of water flowing into the sea, each creating a site of strong scent, it is perfectly reasonable to assume that loggerhead females would approach these sites by following a sort of olfactory gradient to its strongest point. The north-flowing counter current would spread the scent from the Kosi lake system along the beach north of the point of stream entry, which would explain why there tended to be more females nesting there.

This theoretical scent-recognition mechanism would help the turtle in other ways. When a loggerhead emerges onto the beach in the wash zone, it tends to pause for some time. On occasion, in a behaviour known as 'sand-smelling', it thrusts its beak into the sand and sometimes walks along with its beak dug into the sand, leaving a furrow in the beach. It would seem very reasonable to assume that the animal is checking the scent to confirm that its automatic nesting system has chosen the correct site.

Having established that the sand has the correct smell, it then crawls up the beach. It is now proceeding up a 'flare-path' of olfactory cues distributed between the high and low tides every day, twice a day. When it crosses the high tide mark, it is highly likely that these olfactory cues disappear, or are at least much weaker, as a result of the high beach sand being sun-baked and windblown. The sudden change in strength of the cue may be sufficient, again through a process of hard-wired instinct, to inform the turtle that it is safe to dig a nest and that eggs laid here will be safe from inundation by sea water.

Once above the high tide mark, any obstacle – a branch, a pit or a bank – is sufficient to stimulate the turtle to start digging its nesting pit, which is a carefully programmed procedure. Sometimes, while digging their body pit, nesting turtles have been known to thrust their beaks into the newly exposed sand as if once again testing for the presence of the right smell.

To conclude, loggerhead females navigate to the beaches of Maputaland from great distances using their inbuilt GPS system and strand on the beach using an olfactory cue system. As we have many examples of the turtles shifting between nesting events from the south node to the north node, it would seem that they are not wedded to a specific nesting site.

Archie's theory that olfactory cues might play a role in the nesting behaviour of sea turtles has, I believe, been proved correct, despite the fact that the role of the olfactory cue may be operational only in close proximity to nesting beaches and not over the trans-oceanic migrations. His speculative thinking has always been valuable to those scientists following his initial studies.

Leatherback nesting-site selection differs slightly from that of the loggerhead. The leatherback female is much larger, with a soft skin that is vulnerable to damage from rocks and reefs. Leatherbacks thus prefer to use obstruction-free approaches to the beach. As a result, the chosen beach sites tend to receive waves unaffected by reefs and are thus steeper and have much coarser sand. The steep beach means that the female does not have to haul herself too far before finding a suitable place to lay her eggs.

Fortuitously, along the KwaZulu-Natal coast, we find a series of water 'cells', each approximately 700–800 metres in length, caused by the dominant wave patterns, which strike the coast at an angle from the south-east. Each cell builds up water against the beach until it becomes unstable and the surplus water is returned beyond the breakers in what is known locally as a 'rip current'. This is a well-known phenomenon and has caused a great deal of anguish to many unwary or weak bathers over the years. In fact, every year along this coast a significant number of bathers drown as a result of straying into a rip current.

To the leatherback female, however, the presence of the rip current informs her that there is a clear approach to the beach and she uses this current as a guide to her nesting site. They are such powerful swimmers that what is a threat to a human bather is no problem at all to a leatherback. Once she strands, the female labours up the beach with determination and is very difficult to deflect from the nesting process.

I have mentioned that every now and again we find a 'baby in the taxi' event where a female turtle nests on a beach far from the normal nesting area. This is relatively rare, but clearly the female, if under nesting pressure, can use similar cues to access a safe beach site. The fact that they then move to the proper nesting beaches suggests that leatherbacks also have a hard-wired guidance system bringing them back to their natal beaches. Leatherbacks disturbed on their first emergence can move over 40 kilometres in an evening, from one nesting site to another and, in general, display much less loyalty to a section of beach than loggerheads. However, if there is a chemical olfactory cue that is being sought by the leatherback female, it is likely to be somewhat concentrated in the outgoing rip current that carries water that has scoured the beach for hundreds of metres.

Despite the fact that we had carefully mapped the nesting sites of leatherbacks over the years and noted that there was, in most cases, a natural separation between the two species, quite how they managed that had not been apparent to us. It was only when flying back down the coast, when there had been strong southerly winds for a few days followed by an absolutely windless day, that the secret was uncovered. As we flew along the beaches, the rip currents all along the coast were clearly visible as foam-carrying tongues of water shooting out to sea. Only then did we realise that every rip current coincided almost perfectly with well-known and often-recorded leatherback nesting sites. The how and the where of leatherback site selection in Maputaland was solved.

During the 1980s, the beaches of Maputaland were traversed not only by the turtle teams, but also by the security branch of the police, and here and there detachments of military were stationed in case there was an armed invasion. Both security corps were highly sensitive and serious about their responsibilities. On one occasion, a military colonel on holiday at Sodwana Bay came rushing into John Daniel's office (John was the officer-in-charge at the time), demanding the use of a phone to contact Pretoria because 'a huge, tracked amphibious vehicle had landed on the beach during the night!' The colonel had never seen a leatherback nesting track and it took the odd glass of brandy to calm him down.

On another occasion I was called by military intelligence to a special meeting in Kwangwanase to discuss potential landing sites along the coast. The new-to-the-area captain, who was conducting the meeting, charitably said that my knowledge of the coast must be quite good. (After some 20 nesting seasons of walking and driving the coast, I thought this a bit of an understatement.) I was then asked if I could give them a map showing where any incursions from the sea were likely to be successful in avoiding reefs and landing safely.

I immediately assured the captain that I could, but mischievously asked if he was worried about the Russians having the same knowledge because they already had the map. The look of stunned shock on the captain's face will remain with me until I die. He blurted out a question asking how this had happened and I replied that our maps of leatherback nesting sites, published in scientific papers, were as accurate a guide to safe landing sites as one could find. I knew that copies had reached the Soviet Union because I had received letters of acknowledgment of papers from scientists there.

The worried captain did not find this disclosure funny and the meeting ended very quietly. Perhaps he was contemplating charging me with treason.

CHAPTER 15

# Aragonite Blue

The motorcar salesman was very apologetic when he said that the only colour available, in the model of car I wished to buy some years ago, was aragonite blue. To his surprise, I reacted with great enthusiasm (rather like turtle hatchlings, I am attracted to blue things). Apart from the odd mineralogist, I am pretty certain that few people, certainly in South Africa, would have shown any reaction at all to the name. Why was I so excited? Simply because aragonite, a long crystal form of calcium carbonate, is the primary mineral making up the shell of a sea-turtle egg and a sea-turtle egg is a very unique and valuable thing.

Sea turtles place a huge amount of evolutionary faith in the egg and have, over aeons, invested more and more resources in producing as many as possible during their lifetimes in order to ensure the survival of the species. In Maputaland, one of the first aspects of turtle behaviour to which we gave real attention was the egg-laying process. One of the most noticeable, almost unique, features of the sea-turtle egg is that it is soft-shelled. There is, however, a very good reason for this. Sea turtles drop their eggs into a hole some 60–80 centimetres below the cloacal exit point, which means that a hard eggshell would result in a high proportion of broken eggs. So, evolution has modified the mineral structure of the eggshell, providing it with a high percentage of aragonite, which gives the egg a soft, yielding structure that happily permits the egg to bounce off its fellow eggs as it falls into the hole.

One can, in fact, bounce a turtle egg off a hard surface, rather like a ping-pong ball.

Extant sea turtles, which were once land turtles that returned to the sea, have adapted brilliantly to their environment, with perhaps two exceptions: they remain air-breathing animals and they are tied to the land as they have to return to their natal beaches to lay eggs. To this end, as previously described, they undertake lengthy and dangerous migrations and must use this opportunity to maximise their reproductive efforts. Each female has a lifetime supply of ova at maturity and she will ripen as many ova in a season as is possible given her physiological ability. This depends on whether she has amassed enough resources to sustain a long migration and mature a large number of ova.

In Maputaland, the loggerheads are little different from loggerheads in the rest of the world. They lay an average of 116 ping-pong-ball-sized eggs in a single clutch. Each female is capable of laying four clutches during the course of the season, on average, which is a total of some 450 eggs. With 600–700 females nesting in Maputaland per season, the annual production is in the region of 300,000 eggs.

Our leatherbacks, however, tend to lay slightly larger clutches than those in other parts of the world. Our females lay an average of 104 billiard-ball-sized yolked eggs. The term 'yolked eggs' is commonly used because leatherbacks have a widespread habit of laying a large number of yolkless eggs, which generally dribble out after the main batch of yolked eggs. Yolkless eggs come out in a wide variety of shapes, from small and round, through sausage shapes, to long strings of diminutive eggs loosely attached to one another. Yokeless eggs are infertile and result from an over-production of shell material, which generously coats the excess albumen produced to cover the emergent ova in the vitelline duct. These yolkless eggs do not appear to have any function.

An average leatherback female can lay up to 1,000 fertile eggs during the course of the season, which must be an immense strain on the animal. It is commonly believed that sea turtles do not feed while they are nesting, so the energy required for every activity is drawn from their built-up reserves.

With such an effort being put into reproduction, the fertility of the eggs must be high and this has proved to be the case in Maputaland. Very few cases of totally infertile clutches have been recorded. Fertility rates of up to 90 per cent are common in loggerheads and the figure is only slightly less in leatherbacks. Loggerheads outnumber leatherbacks by a factor of nearly ten in Maputaland, and are clearly more successful, but the modest difference in fertility does not explain the difference in hatching success. To explain the lower hatching success of leatherbacks, one must take into account where each species lays its eggs.

Loggerheads, being lighter and more nimble, tend to nest higher up on the beaches, often above the primary dune and sometimes far above it. One famous female managed to climb the secondary dune to a height of nearly 100 metres, assisted by the fact that she happened to ascend a section of beach that had its upper vegetation cleared by the wind. At the sites chosen, nearly all well above the high-tide mark, loggerhead nests have little chance of being inundated by sea water, as the eggs are in well-drained sand. The only dangers facing such clutches are posed by ants and other predators, and the wind.

Turtle females expend a huge amount of energy preparing a nesting site. They will often begin to dig a body pit, find the site unsatisfactory (we have yet to find a sound reason why), and start again elsewhere. Some female loggerheads have been known to try ten different sites before settling down to complete their task. The typical female uses her fore-flippers to excavate a body pit, which brings her body below the surface of the beach. This removes the dry, friable surface sand and exposes the damper, more stable, sub-surface sand, which will hold the shape of the nest without collapsing. On the silica sand beaches of Maputaland this is not a major problem. On tropical islands such as Europa, however, where green turtles are faced with much finer coral sand, which is dried out to a greater depth by a stronger sun, they have to remove huge quantities of surface sand before finding suitable damp sand.

The hind-flippers are used alternately as the female excavates the egg hole. Each flipper scoops out sand, which is then deposited on the side of the hole. The opposite flipper then repeats the process, bringing out another cupful of sand. The previous flipper now flicks forward sharply, throwing the sand previously excavated over in front of the turtle, thus keeping the level of the hind-flippers the same throughout the digging process. The sand dug out is flicked forward and away from the nest hole to prevent accidental infilling. The turtle completes the process by reaching as deep as she can, often lifting her body with her fore-flippers, and scraping the edges of the hole, ultimately creating a flask-shaped depression that is sufficiently large to hold her 100 plus eggs.

Loggerheads will then straddle the hole, each hind-flipper firmly planted on either side. Leatherback females seem a little coy and always hang one hind-flipper into the hole, it can be the left or the right, as if to ensure that no sand falls into the clutch. Both species pause briefly at this point and then with some effort begin to lay their eggs in bursts of up to four at a time. The laying of a full clutch can take up to half an hour.

Once the eggs are laid, both species begin to scoop sand into the hole with their hind-flippers. Each flipperful is followed by an amazingly delicate probe with the flipper to feel the depth of the sand. This process is repeated until the sand deposited has reached the desired level. The females then press down firmly with the hind-flippers and, in so doing, form a compressed crust of sand above the eggs. Some females exert considerable pressure to create this crust, rising up on their fore-flippers to use their body weight.

Thereafter, all turtle species use the fore-flippers to hurl sand over the nest site and may move many cubic metres of sand in doing so. Leatherbacks have been recorded spending two hours or more, disturbing over 100 square metres of beach, in disguising their nest site.

Maputaland beaches are scoured by the winds and, in one way, this is a great help to the nesting turtles as a brisk wind will rapidly remove all trace of a turtle's visit and her nest site. During a southerly 'buster', the amount of sand transported by the wind is astonishing. Happily, because loggerhead eggs are normally situated at least 60–80 centimetres below the surface of the beach, most are protected from all but the very worst of windstorms. Even then only a very few nests are exposed and destroyed. After a good blow, however, it is very common to find clutches of eggshells, from nests long hatched and gone, appearing on the surface of the beach.

Sand deposited by the wind can make things difficult for emerging hatchlings. On one occasion, when trying to find the remains of a clutch that had emerged, we had to dig down 3 metres before we found the nest. It was so deep that one of my colleagues eventually had to hold me by the ankles as I disappeared deeper and deeper into the sand. The hatchlings must have been very weary after struggling through that sand to reach the surface. It must have taken days.

Leatherbacks tend to nest much lower down the beaches on the face of the primary dune. Having a much longer reach with their hind-flippers, the leatherback females dig much deeper nests. If they have nested a little too low on the beach, they expose the eggs to the possibility of inundation by salt water due to high tides and storm surges. If there is too much water in the sand surrounding the eggs, there is a danger that air will not be able to percolate into the egg cavity and the emerging hatchlings literally asphyxiate.

Depending on the ambient temperature during a season, loggerhead turtle eggs in Maputaland take about 55–65 days to hatch, and leatherback eggs, because it is cooler at the greater depths of their nests, can take up to 75 days. During this period under the sand, wondrous things happen to the egg.

A marked difference between bird eggs and sea-turtle eggs is that, after some 24 hours, the developing turtle embryo attaches to the inner surface of the shell. This is invaluable, as the developing embryo extracts minerals from the eggshell and incorporates them, which reduces the physical strength of the shell. Once the youngsters are ready to hatch, they have to break their way out of the eggshell, which has become thin and brittle, with a small 'egg tooth' on the beak. The role of the egg has ended.

Unfortunately, the egg is the target of many terrestrial predators. The worst of all, perhaps, is man. Throughout the world, wherever sea turtles have gathered to lay eggs, human beings have benefited from

# FIVE SPECIES OF SEA TURTLE

1. **LEATHERBACK TURTLE** (*Dermochelys coriacea*)
2. **OLIVE RIDLEY TURTLE** (*Lepidochelys olivacea*)
3. **GREEN TURTLE** (*Chelonia mydas*)
4. **HAWKSBILL TURTLE** (*Eretmochelys imbricata*)
5. **LOGGERHEAD TURTLE** (*Caretta caretta*)

Designer: Di Martin. Reproduced courtesy uShaka Sea World.

6

7

8

9

ANCORA
CONTAINS PIECES OF
SELECTED TURTLE MEAT
3 TO 4 SERVINGS
"The King of Soups"
NET WT. 1 LB. 4 OZ.
CLEAR GREEN
TURTLE SOUP
WITH IMPORTED SHERRY WINE
CLEAR GREEN TURTLE SOUP
WITH IMPORTED SHERRY WINE
INGREDIENTS: GREEN TURTLE MEAT FROM DEEP SEA CARIBBE
GREEN TURTLES, IMPORTED SHERRY WINE, ONIONS, CELERY, CARROT
TOMATOES, MONOSODIUM GLUTAMATE, SPICES, SALT, AND ENOUGH WATER
PREPARE PROPERLY.
DIRECTIONS
TO SERVE HOT
Empty contents in a saucepan and bring to boiling point. Stir to preve
scorching. Sherry may be added if desired.
TO SERVE JELLIED
Empty contents in cups and place in refrigerator for about 4 hours. Sher
may be added, if desired, before soup is placed in refrigerator to be jelle
BON VIVANT
Soups, Inc.
NEWARK 7, N.J. * U.S.A.

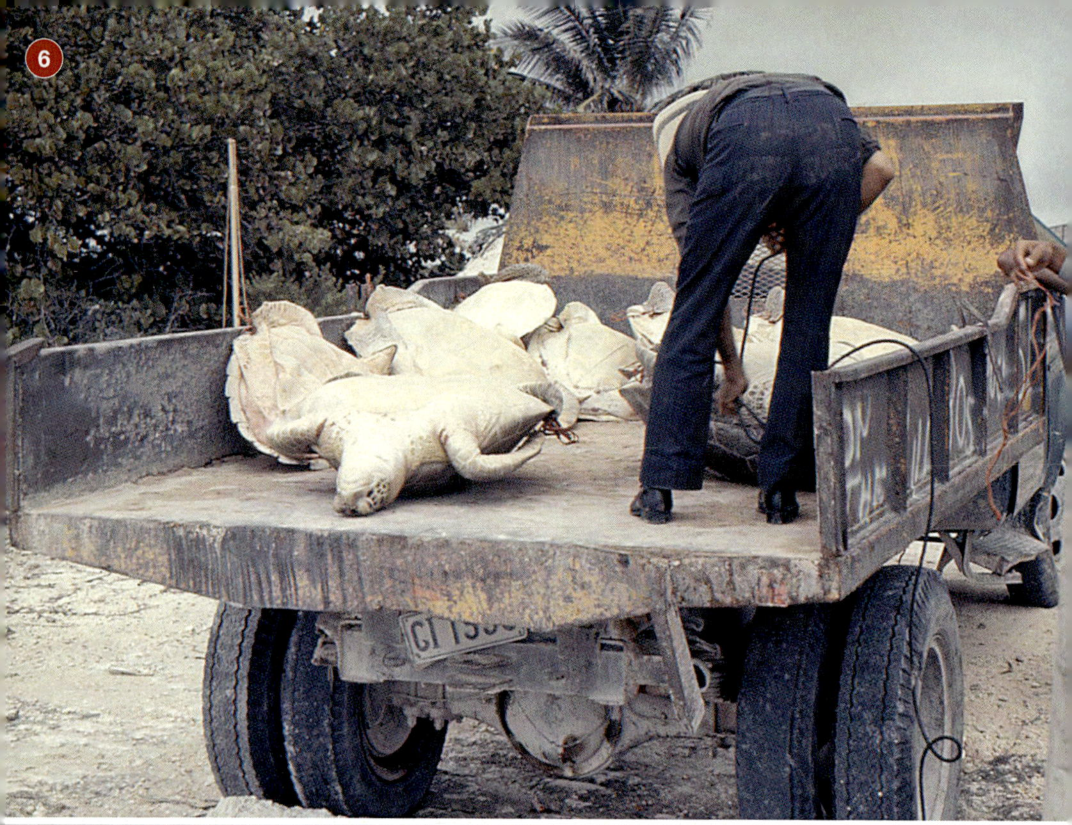

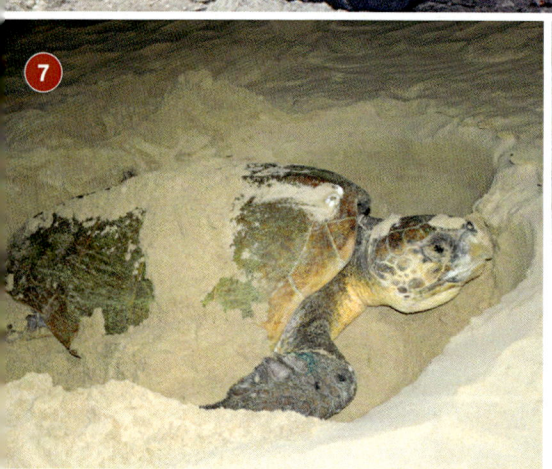

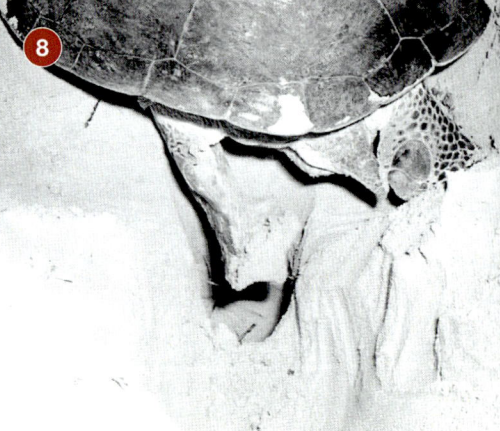

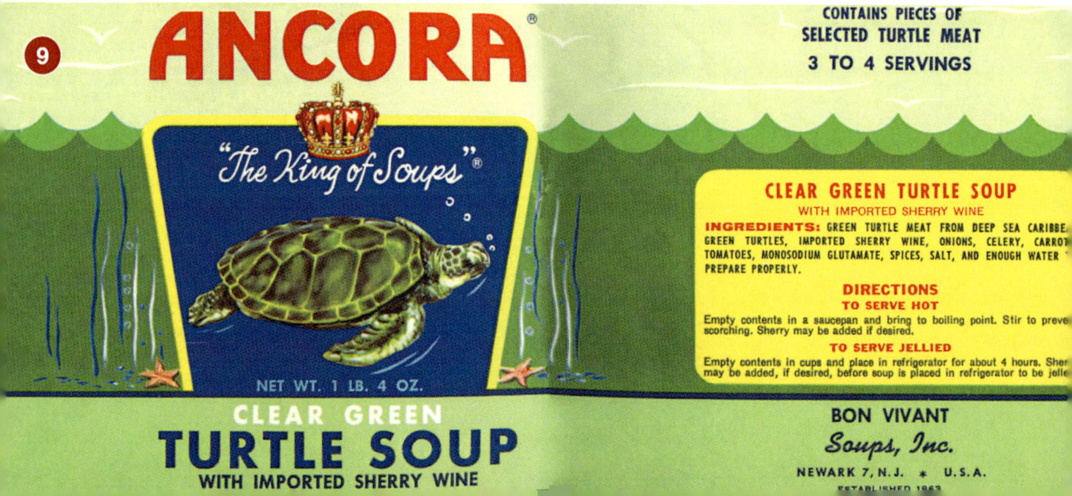

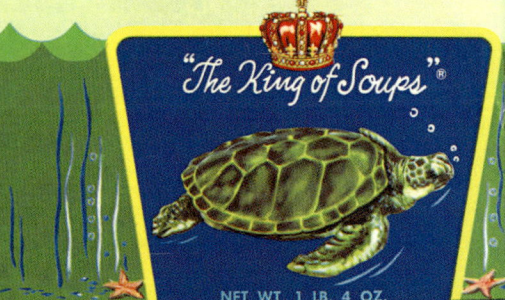

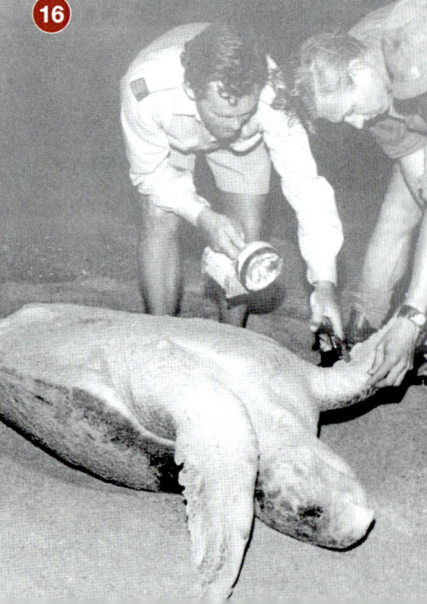

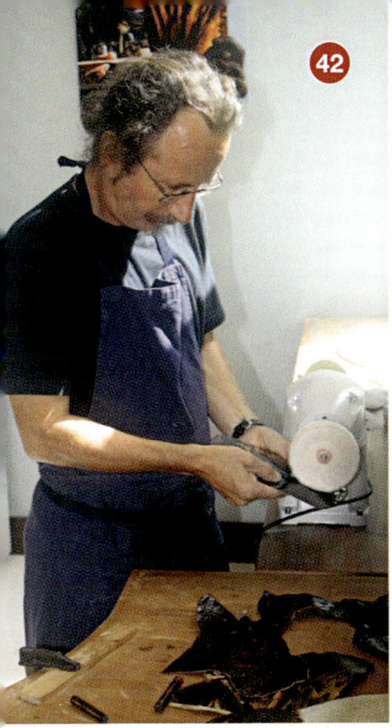

46

47

55

56

57

58

59

this bounty of nature. For the most part, where this was merely local people taking what they needed to survive, no problems were posed to the survival of the turtles. Once the exploitation became commercial, however, severe problems soon became apparent. One of the most famous cases of overexploitation of eggs occurred in Terengganu, Malaysia. In the 1950s, as far as our knowledge at the time was concerned, it had the world's largest colony of leatherbacks.

In many Muslim countries adult turtles are not killed and eaten because they are regarded as unclean. There are no such restrictions on the eggs, however, and almost all Muslim countries have exploited their turtle-egg resources, and continue to do so in many places. Despite the heroic efforts of some of our Malaysian colleagues, who brought all sorts of conservation programmes to bear on the problem, the leatherbacks of Terengganu are effectively extinct. Local turtle conservationists tried everything from licensing egg collectors, to buying eggs from the collectors and placing them in extensive artificial hatcheries. Their efforts were all in vain.

Tom Harrison tried to manage the traditional egg harvest in the Talang Talang Islands off the coast of Sarawak without success. Quite literally millions of green-turtle eggs were collected every summer, terminating in one special festival holiday on which it became a local tradition to have egg fights. Thousands of perfectly good eggs were destroyed in an orgy of egg fighting, not unlike the incredibly wasteful tomato holiday in present-day Spain, where tons of tomatoes are similarly destroyed.

In Mexico, egg collecting for sale in the cities brought the Kemp's Ridley turtle to the brink of extinction, as it was popularly believed that their eggs had aphrodisiacal properties. The same belief was widespread on the Pacific coast of Mexico. Along with the slaughter of adult olive Ridley turtles for their leather, the illegal removal of eggs from the surviving nesting beaches caused serious population crashes. The Mexicans eventually brought in the army to restore control and I am happy to report that they achieved miracles. With the growing awareness of the Mexican government, which has supported some outstanding biologists such as Drs Rene Marquez and Georgita Ruiz, the conservation situation in Mexico on both the Pacific and the Gulf coasts is rapidly improving.

The situation was only slightly different along the Maputaland coast. Although the killing of adults was not that common until after the Second World War, nest robbing was widespread. In the Sodwana Bay section, virtually every clutch of eggs was robbed the morning following its laying. The eggs, for the most part, were taken for subsistence use. Nevertheless, this intensive egg harvest was probably why adult numbers were so low and why the presence of turtles along that coast had escaped discovery for so long. Happily, with the start of the turtle-protection programme, all such collections were drastically curtailed and were eventually stopped completely, which was reflected in the growth in populations of nesting adults over the 25 years that followed.

Of course, just when we thought that the problem was under control, disaster struck in the form of HIV/Aids. When the epidemic reached serious proportions at the end of the twentieth century, it did not take long for some charlatan of a sangoma to come up with the idea that eating sea-turtle eggs cured HIV/Aids. There was a spurt of egg poaching, but with the cooperation of the local Thonga *amaKhosi* and the staff of Ezemvelo KZN Wildlife the problem was controlled and seems to have disappeared.

Back in the 1960s, the local amaThonga used to feed turtle eggs to their chickens because they believed it made them better egg layers. Personally, I feel that the Thonga chickens have a pretty hard life and any improvement in protein intake, whether supplied by turtle eggs or not, would have improved their egg-laying capabilities.

In my experience, turtle eggs are overrated both as a food and as an aphrodisiac. During the course of my research, I have had the opportunity of trying both loggerhead and green-turtle eggs. This was, of course, in a spirit of open-mindedness and not wishing to waste research samples. Loggerhead eggs are simply not tasty, despite the fact that, according to Archie Carr, local Florida residents in the United States believed that they were better than chicken eggs when it came to baking cakes. You can fry turtle eggs, of course, but they are a challenging breakfast. The albumen of a turtle egg does not turn white when fried and it takes a certain amount of desperation to eat an egg that does not appear cooked. I have never managed to finish one.

Perhaps that is why I never noticed any post-breakfast sexual excitement.

On Europa Island in 1970, where the meteorology team was supported by an excellent Creole cook from Reunion, my green-turtle egg samples were skilfully turned into a scrambled-egg concoction with spices, which was, to be honest, delicious. There was a general cry of anguish when my sampling was finished, which suggested that the broad consumption of turtle eggs by the Reunionese was not entirely a thing of the past. Today, even the pressure of the odd meteorologist cook has gone, as all weather recordings on Europa are done automatically and remotely.

Protection of the sea-turtle egg is now recognised as an indispensable component of successful sea-turtle conservation. The eggs have the advantage of being able to re-establish spent colonies. In many countries, millions of eggs have been translocated to hatcheries to reduce the pressures of poaching and to protect them from overexploitation by predators. Good conservation practice, however, recognises that there must be demonstrable reasons for removing eggs from natural beaches to hatcheries. Natural beaches with good populations of turtles and few threats to a naturally laid nest should not be interfered with.

A final comment may be worth making. Over the past few years, a series of well-meant, but scurrilous, emails containing images of egg-collecting along the Pacific coast of Costa Rica has been widely distributed. The emails were accompanied by negative remarks suggesting that this practice is the cause of the disappearance of turtles. This is regrettable because the source of the images is a sustainable-use programme developed by the conservation authorities and scientists of Costa Rica after years of research. The programme required a major confidence-building endeavour involving the local communities that traditionally exploited turtle eggs. They were stopped from doing so some years ago by the Costa Rican government in its pioneering efforts to establish a good turtle conservation regime.

Costa Rica has an excellent record of turtle conservation and, in 1956, was one of the first countries in the world to create a turtle sanctuary. This was the site of the famous green-turtle beaches of Tortuguero on the Atlantic coast, where Archie Carr and the Caribbean Conservation Corporation inspired by him have done outstanding work. This has resulted in a staggering increase in the green-turtle populations nesting there.

The eggs being harvested by the local community are those of the olive Ridley turtle. Along this part of the Pacific coast, these turtles tend to nest in arribadas over several days. Millions of eggs are destroyed by the huge numbers of nesting turtles that pour up the beaches in their tens of thousands. It has been calculated that on some of their beaches only 0.2 to 5 per cent of hatchlings emerge after surviving this early onslaught. It thus made good sense for the Costa Rican authorities to permit the traditional collection of eggs on the first two days of an arribada and thereafter the villagers assist in the protection of nesting females. This is a practical and sensible programme, which deserves the praise of the world and not the condemnation implied by the circulating emails.

Sea-turtle eggs are rarely wasted in such exceptional circumstances and that is a good thing, in my opinion. Protection of turtle eggs is an absolute necessity and as important as the conservation of adult females. Selfishly, perhaps, I believe that there is another good reason for protection. Without the evolutionary heritage of the egg, which has bound sea turtles to the beaches, turtle researchers and conservationists around the globe would never have had the hands-on joy of working directly with these magnificent reptiles.

No one could have been more pleased than me to own a car that was aragonite blue.

CHAPTER 16

# Politics and Turtles

Sea-turtle conservation is an apolitical activity and should be far above the sordid machinations of politics. We realised that this was not entirely the case when the odd hand grenade arrived, but we were convinced that this was a brief intrusion. Expectations are, alas, seldom met in the real world. The first major global impact we experienced was after the Israeli Six-Day War in 1967. General Nasser, in a fit of pique, blocked the Suez Canal, necessitating all oil tankers, both responsible and irresponsible, to travel around the Cape of Good Hope to serve their markets in Europe and the Americas. Shortly afterwards, every fly-by-night oil tanker appeared to use our coast to clean out their tanks before returning to the oilfields. As a result, the turtle beaches were covered in oil – never enough to cause a major disaster, but enough to necessitate supplies of paraffin so that the turtle-trotters could remove all the oil adhering to their legs. We had sandy oil deposits on everything – vehicles, clothes and tagging tools. The sticky mess even found its way into our bed sheets.

This was a constant problem for some three years and a great cheer went up when the Suez Canal was reopened for business. The oily remains of the slicks were visible for years, however, even though they had dried out and were buried in the beach. We were, of course, concerned about the impact that these minor slicks and so-called tar balls had on the turtles, both hatchling and adult. Happily, we never observed any turtles, damaged or dead, whose condition could be blamed on the oil. More recent events in the Gulf of Mexico, with the explosion of the BP oil platform, demonstrated that massive oil spills can smother hatchlings and harm adult turtles. We were extremely lucky that our political oil spills never attained such serious proportions.

A more localised political event, however, had quite genuine repercussions. It launched a decade-long programme and probably cost the Natal Parks Board and the KwaZulu Bureau of Natural Resources, its partner at the time, a small fortune in staff time and vehicle and fuel costs. It certainly placed a huge burden on the local staff, both those directly on the turtle teams and those at Sodwana Bay.

The first I knew of this impending disaster was when John Geddes-Page, the director of the Natal Parks Board, burst into my office shouting, 'What are you going to do about it?' This occurred in 1982, when I had just assumed the role of assistant director, conservation, in the Natal Parks Board, which brought me into much closer association with our volatile and excitable director. On asking what 'it' was, he informed me, almost shaking with rage, that the 'bloody government' had just decided to give Swaziland a piece of Maputaland, which would allow the Swazis access to the sea. What was worse, it included the entire Kosi Bay estuary that was to be developed into a deep-water harbour for Swaziland's exports. I then joined John in a general cursing session that would have reduced our esteemed minister of Foreign Affairs to tears.

The strip of Maputaland included one of our most famous and best-loved game reserves, Ndumu, as well as the Tembe Elephant Park, then being planned by our KwaZulu colleagues, as well as the Kosi Bay Nature Reserve and the estuary system that we all hoped to see protected one day. Apart from this, however, the 'gift' also included some 25 kilometres of the best turtle-nesting beaches under our protection. I nearly had a fit, as the proposed mouth of the harbour would exit right in the middle of our most dense area of loggerhead nesting sites, right between the two target nodes of our growing populations.

This cheap political charade was announced without consultation and completely ignored the fact that, with the enthusiastic support of the national Department of Sea Fisheries, the first marine reserve in Zululand had recently been established slightly further down the coast. All concerned were in the throes of seeing the northern beaches right up to the Mozambique border declared as a marine reserve, but now they were to be given away. The government also ignored the fact that the Kosi Bay estuary, the turtle-nesting beaches and the coral reefs offshore had received international recognition through their registration as a Ramsar site of global significance.

The general popularity of the national government reached a new low in nature conservation circles in Natal. Prince Mangosuthu Buthelezi, the chief minister of the KwaZulu government, was enraged, and not without cause, as the government intended to give away a large piece of the Zulu kingdom.

As insurrection was not an option, my colleagues and I immediately undertook a survey of the beaches in the new St Lucia Marine Reserve, south of Sodwana Bay, in order to find a suitable stretch of beach to which we might translocate turtle eggs. This was a drastic move, but the excavation of the mouth alone would virtually destroy our prime nesting sites and, in our experience, the government was unlikely to change its decision.

This task was not as simple as it might seem. Turtles, especially leatherbacks, commonly nested on the beaches south of Sodwana, but this was a tiny fraction in comparison with the Kosi beaches. The most densely used beaches were in front of Lake Hlange, the biggest of the Kosi lakes, and a long string of large pans slightly further south. The presence of a significant body of water near the beach was a prerequisite, as we believed that water emerging under the beaches from these bodies of water may have been providing an olfactory guide to returning turtles (see Chapter 14).

In addition, we had to ensure that the beach chosen was warm and received day-long sunshine. The sex of turtle hatchlings is determined on about the 19th day of incubation and depends on whether the temperature of the incubating clutch is high or low. (We know this thanks to the excellent research of Dr Nicholas Mrosovsky, of McGill University, Canada, and his colleagues.) High temperatures produce females and low temperatures produce males. The so-called pivotal temperature, in the case of the Maputaland beaches, was around 28.7° Celsius. (We know this thanks to the work of Jenny Maxwell, then of the University of Durban, Westville. This has been confirmed more recently at around 29.2° Celsius as a result of a study by MK Boonzaaier of the Nelson Mandela Metropolitan University.) As there was little point in choosing a beach that would remain cool throughout the season, and thus produce only males, we had to find a beach near a freshwater source that was exposed to the sun for most of the day. After checking all the way down to Cape Vidal, we selected a beach near the inland freshwater lake called Bhangazi North, which was nearly 100 kilometres away.

If we were going to undertake this endeavour it would have to be on a worthwhile scale. We had run a small local hatchery at Bangha Nek, involving less than 1,000 eggs each year, for about three years. This was more for experimental reasons than any conservation need, but the results had been quite satisfactory. The proposed translocation, however, was on a scale that we had never considered before. We decided that we would translocate 20,000 eggs per season for as long as was necessary.

Our experience, both in our hatcheries and those in other areas, had shown that there were severe constraints involved in the translocation of eggs. They had to be moved as soon as possible after laying, preferably within 12 hours. The reason for this was, as mentioned before, that the developing embryo very quickly attaches to the 'roof' of the egg. If an egg was moved after the embryo had attached, any movement could cause the yolk to tear away. If this happened the embryo would die. As most eggs are fertilised before laying, we had no way of knowing the amount of time that elapsed between the fertilisation and the laying of the eggs. A severe southerly gale, a lightning storm or heavy rain were not favoured by nesting loggerheads, which did not emerge to lay on such nights. Studies had shown that these delays did not appear to bother the females, so they could clearly retain a clutch for some days after fertilisation. What was not known was whether the general movement in the oviduct prevented embryo attachment before the eggs were laid. It was reasoned that the attachment does not happen inside the oviduct because the shock of an 80-centimetre drop into the nest would surely have torn the tissues. We were, however, unsure of all this, so our primary goal was to transfer the eggs with minimal movement.

A second constraint was that turtle eggs do not respond well to temperature shock. So, any transfer during the heat of the day, when the eggs might suddenly be exposed to bright sunlight, was out of the question. All movement of eggs would have to take place on the night of collection.

Yet another constraint was the fear of contamination of the eggs. Contamination may have occurred if the collector's hands were dirty or if he or she was careless and broke an egg – the leaking contents would act as a beacon to predators such as ants and ghost crabs. Larger predators were less of a problem because the hatchery site was to be securely fenced and each clutch surrounded by bird wire.

Finally, if the beaches in the north had a strong olfactory cue used by the nesting females, it was essential that we include a decent-sized sample of the sand that surrounded the eggs in the original location. For all we knew, the cue may have been left by the nesting females themselves, as the eggs are accompanied by copious amounts of mucous. It was considered possible that, after thousands of years of using the beaches, nesting females left a stable pheromone or chemical that informed future generations that the beach was a safe place to nest. It was also possible that the breakdown products of the developing hatchlings, and the eggshells remaining after the clutch hatched, were the source of a stable chemical indicator. In all likelihood, it was a combination of these possibilities. It was clearly imperative to include a large sample of Kosi beach sand in the lining of the artificial nest in the new hatchery at Bhangazi North.

Once the decision had been made to proceed with translocations, a number of very large cool boxes were purchased and shipped to Bhanga Nek. An open-mesh plastic 'bucket' was designed so that the eggs had to be handled only once on collection from the natural nest. Once in the new transport bucket, the entire clutch could be placed into a Styrofoam box on a bed of sand excavated from the nest. When they were in the box, additional sand from the nest was poured around the bucket to cushion it from the inevitable movement as the vehicle travelled south along the beach. The Styrofoam boxes, which had ventilation holes drilled through the top, were then closed, sealed and collected to wait until a full load had been prepared.

When the tide was right, the load was transported down the beach to the new site, which had been prepared by the staff under the supervision of Mike Bouwer at Sodwana Bay. The hatchery had been fenced and suitably signposted with an explanation of the purpose of the exercise. This would hopefully discourage tourists and others from entering and interfering with the newly transported clutches. The good behaviour of our tourists is a matter of some pride – in the whole ten years of the exercise, we were not aware of any tourists entering the enclosure, even though it was unguarded.

On arrival, the staff excavated new nest holes to the appropriate depth, shaped the nest using clean hands, lined the cavity with sand from the Styrofoam box and then the entire bucket of eggs was lifted out and placed carefully into the new nest. All the sand from the original site was then poured into the cavity to cover the eggs. The plastic bucket remained in the nest until the hatchlings emerged. After the first season, we had to redesign the plastic bucket using a smaller mesh, because we found that some of the hatchlings managed to get their heads stuck in it. They were rescued when the hatched nests were dug up by staff in order to ascertain the hatching success of each clutch.

Over the next ten years, until 1992 when the whole foolish idea of the land transfer to Swaziland was forgotten, we transported 200,000 eggs to their new home. We enjoyed a hatching success rate of 67.7 per cent, so the care and attention given to the eggs and the care taken in transport were worthwhile. We never marked any of the hatchlings emerging from the hatchery, as we wanted to minimise any possible chance of injury or disturbance to them. We could never be certain of the success of the endeavour because that stretch of beach is not patrolled, or nests counted, with the same intensity as the main turtle beaches 100 kilometres to the north.

If the attraction of the northern beaches proved to be too strong, then the mature females in years to come would simply go back there and no harm would have been done. If the port development had gone ahead as planned, then perhaps the signals further south, even had they been much weaker, may have saved the day. In the absence of any olfactory cue in the north, due to the negative impacts and pollution from the port, we hoped that turtles would be brought back to the area by their imprinted GPS guide. It was a long shot, but worth the effort.

The question remains why the government decided to give away a significant part of South Africa. The proposal motivated the KwaZulu government to launch a massive campaign that changed the political atmosphere of Maputaland forever.

The national government managed to keep the real reasons for their machinations a closely guarded secret, but three apocryphal tales to explain these events have emerged over the years, the last story appearing in print in 2012.

This most recent tale suggests that the Seychellois were threatening to kick the Americans out of their airbase on the Chagos Islands so that the original communities could be resettled in their ancestral homes. The Americans would then have to build a new base in the Indian Ocean. As they could not deal openly with the fast-becoming-a-pariah state of South Africa, the new base at Kosi Bay could be built only if it was deemed to belong to Swaziland – an independent state requiring development assistance. South Africa, it was suggested, went along with this entire concept. Very commendable, no doubt.

The second tale has it that 'the South African government decided to give Swaziland the magisterial district of Ingwavuma (which includes Kosi Bay) and the Swazi "homeland" of Kangwane in return for Swaziland's cooperation in ousting the African National Congress from within its borders and for various undisclosed political favours that would make South Africa less vulnerable in its fight against terrorism' (Mountain, 1990).[1]

My favourite tale is much more romantic, however. In 1981 a bunch of mercenaries calling themselves the 'Ancient Order of Froth Blowers' were engaged, by persons unknown, to undertake a coup in the Seychelles, overthrow the socialist president and install a more conservative government. Not wishing to engage a South African Airways flight, these stalwart defenders of democracy used a Swazi Airlines plane. With the usual efficiency we have come to expect from such derring-do, they arrived, safely but probably inebriated, at Mahé airport to be accorded a less than welcoming reception by the Seychelles military. The Seychellois, responding with understandable enthusiasm, forced the surrender of many of the mercenaries by spraying the Swazi aeroplane with bullets, rendering it, and a few of its occupants, *hors de combat*. The coup was a spectacular and embarrassing failure, despite the fact that some of the would-be warriors hijacked another plane and escaped.

Rumour has it that, shortly thereafter, the Swazi government, understandably peeved about the loss of about 25 per cent of its commercial air fleet, and clearly aware that the South African government was deeply involved in the entire disaster, threatened to spill the beans to the world at large. This would have depressed South Africa's failing public relations programme even further. So, the South African government, caught with its pants down, bowed to the genteel blackmail and agreed to give Swaziland a route to the sea, for, of course, the noblest of reasons.

I have no idea which of these tales, if any, has any basis in fact. It is a matter of fact, however, that the offer was made, caused the conservation and political communities of Natal and Zululand immense anguish and cost the Natal Parks Board inestimable thousands of rands. The third version, the one that attracts me the most, is just the sort of ill-fated scheme that one might have expected from the securocrats at that time in South Africa's history. The fact that it was led by the 'Ancient Order of Froth Blowers' gives it a special place as one of history's greatest fiascos.

1 Mountain, A. 1990. *Paradise Under Pressure*. Johannesburg: Southern Book Publishers.

CHAPTER 17

# Conferences, Politics and Astonishment

While turtle conservation is apolitical, global politics often takes precedence over the goals of a conference and can have both painful and alarming effects on scientists and conservationists attending such events. My own experiences have shown, however, that the peer respect and friendship of fellow biologists can win through in the face of negative global politics.

By 1971 turtle biology had a death grip on me and I was more than thrilled when Allan Heydorn at ORI suggested that I should represent the institute at the first Biology of the Indian Ocean conference in Kiel, Germany. Several other South African biologists and marine physicists would be going, so I would be in good company (well, it would be a pity to disillusion Allan now, so I shall stick to the relevant facts). Drs Butch Hulley and Brian Kensley from the South African Museum in Cape Town and Dr Johann Lutjeharms from the Department of Sea Fisheries were my fellow attendees. Butch eventually became director of the museum and Brian ended his career with the Smithsonian Institution in Washington DC. Dr Lutjeharms and I had little to do with one another in Kiel, but I have since contributed to joint papers on sea turtles with him, calling on his remarkable knowledge of the Agulhas Current. In Kiel, however, Butch, Brian and I formed what might be termed an unholy trio.

When it comes to organisation, the Germans are in a class

of their own and I believe that the Kiel conference was one of the most enjoyable and hospitable that I have ever attended. Our hosts did everything to make the conference memorable. It was the only conference that I have attended where a speaker was bodily carried off the stage by two seriously large German students after he exceeded his time limit. Not to be deprived of his moment of glory, the scientist dragged the microphone with him and his voice was still delivering his contribution long after all concerned had disappeared from view. The sudden squawk of the microphone being wrenched away brought a cheer from the audience and the chairman had some difficulty in maintaining order.

The conference was also memorable for the incredible series of after-session parties. They were great experiences and provided the chance to meet scientists from all over the world. The most memorable participants were the Russians, who were hardly fans of South Africa. Brian, Butch and I met the Russian delegation at the third party we went to. Their leader was a marine biologist, Dr FA Pasternak, nephew of the famous author Boris. Dr Pasternak was a very large man with a very large presence. His colleague, Prof. Anna Lisitzin, was also a marine biologist. Wherever they went, as was the practice in those days with Russian delegations, they were accompanied by their political officer, who was a small round lady. When coincidence brought us into close proximity at the party, I was determined that we should meet them. After all, they were the *Rooi Gevaar*.

Throwing caution to the wind, I went over to Dr Pasternak and introduced the three of us. He peered down at us suspiciously and said, 'Wher-r-re do you come from?' I replied that we were from South Africa, to which he replied, 'I know Durban and Cape Town. Beautiful country. . . terrible government!' Mildly stung by this observation, I asked him where he came from, to which he replied in surprise 'Rooshia!' I immediately replied, 'Ah yes. Beautiful country, terrible government!' All three Russians looked at us and then burst out laughing and said, 'Come, we drink vodka!' These two unlikely groups became virtually inseparable for the rest of the conference. My fondest memory of this *entente cordiale* was the sight of Anna, at the end of the final Kiel City banquet, wading through the masses waving a small badge of the Soviet Academy of Sciences, which she insisted I take as a souvenir. I still have it to this day.

The political officer, whose name escapes me, eventually became very warm and friendly and, in a frank moment, said that she had good friends in Johannesburg. I immediately offered to pop in and see them the next time I was in the area to give them her regards. She looked at me oddly and said, 'You must be joking!'

Not all conferences were such a joy, however.

In 1989 a conference was organised in San José, Costa Rica, and my United States colleagues invited me as a guest speaker. The Western Atlantic Turtle Symposium was aimed at providing the best available experience of turtle-conservation techniques from across the globe to the Caribbean, Central American and South American countries with nesting sea turtle populations. By that time, our programmes in South Africa had received a great deal of praise and, to Dr Fred Berry and the other organisers, my presence was seen as an asset. I felt deeply honoured to have been invited and was somewhat flattered to have Fred note my arrival at the first morning plenary session. At the first opportunity he announced that the gathering was now complete as the last guest speaker, Dr George Hughes, had arrived from South Africa. Little did I realise just what an impact that announcement had made.

It was an intergovernmental conference and many senior civil servants and a few ministers were attending as leaders of delegations, supported by their biologists and conservation staff. The mention of South Africa, thanks to its adherence to apartheid, was like a red rag to a bull and apparently sparked a series of meetings among the Caribbean states that were aimed at boycotting the conference. When my turn came to give my presentation, I, being a nervous speaker at the best of times, hardly noticed that the audience had shrunk dramatically. When I finished, the chairman of the session, and a friend of note, Dr Peter Pritchard, told me that every Caribbean delegation had walked out, along with those of a number of other countries. They were gathering in another breakaway room (I had often wondered where that name originated and now I knew) with the stated intention of finalising a planned mass withdrawal. To say I was horrified would be an understatement.

I have, since leaving school, had an open dislike of apartheid and have practised what I preached throughout my career. As a result, I have endured some fairly derogatory attacks for my views. The thought of the symposium failing was, therefore, very distressing, especially as so many of my colleagues had put so much effort into its organisation. So, I asked where the meeting was being held and immediately went to the room and walked in. Needless to say, I did not receive a standing ovation. In fact, my appearance led to an immediate silence. I have often heard talk of silences in which one could expect to hear a pin drop. Well, this was one of them. My heart was pounding like a traction engine as I requested an opportunity to address the gathering.

Out of sheer shock, I think, permission was granted and I took the microphone. I explained that I fully understood their views, but would be disappointed if they took out their distress on the conference. I pointed out that all the turtle biologists there, including myself, had been invited because we had experience in turtle conservation techniques and we had come to share our knowledge with the attendees. I told them that I considered it a grave disservice to my colleagues that their efforts should be negated because of my presence. I said that I would, consequently, remain out of the plenary and other sessions and would remove myself from the conference entirely, if they deemed it necessary. With that, I thanked the gathering for its courtesy and walked out of the room, accompanied by a deathly silence.

It is no small thing to be potentially responsible for the collapse of a very large conference. I was shaken and took refuge in a hall of the hotel where all the posters were on display. I nervously wandered about trying to look calm when into my consciousness came a broad Australian accent shouting, 'What the bloody hell is going on? Who started this crap? I'll go and knock the shit out of him!' Suddenly Col Limpus from Queensland, a friend of nearly 20 years, appeared next to me. (He has since become probably the finest sea-turtle biologist in the history of sea-turtle science.) He threw his arms around me and said that he was prepared to defend me and my honour, physically, if necessary, by speaking publicly on my behalf. 'This is bullshit!' he cried.

What else could I do but offer to buy him a beer to calm him down?

Later, a Caribbean spokesman came to me quietly and said that, as long as I did not speak publicly again, I was free to attend all sessions of the conference without repercussions. The threatened withdrawal had been annulled. Fred and his colleagues (and, need I add, myself), were more than relieved and the conference was eventually a great success. What really was the greatest gesture of all, however, was that my conservation and biologist colleagues from Mexico and the Caribbean, the leading countries in the insurrection, invited me to sit at their table at the final banquet following the closure of the conference. This was in defiance of their political seniors and that is what real courage is about. I shall never forget them and the debt I owe them.

One seldom attends conferences at which lawsuits are threatened and general chaos results, but the first World Conference on Sea Turtle Conservation, held in 1979 in the State Department building in Washington DC, was just such an event. During the previous ten years there had been a great flowering of NGOs dedicated to any number of conservation causes, and the declaration of CITES in 1973 had created an outlet for all their passions. The media, of all sorts, are eager for stories that are out of the ordinary, especially those that have a standpoint at odds with the establishment. NGOs fill this need admirably because their passions about their issue, or issues, often provoke them into making statements that are a little deviant from the truth and, if controversy is established, the media encourage them with enthusiasm.

It is not that NGOs often have a view different from that of the formal authorities, and there are certainly occasions where their views are more than vindicated. The problem lies more in the fact that many NGOs are *single-issue* conservationists and have neither the interest, nor the will, to consider other points of view. Sea turtles have attracted a huge number of enthusiastic people who care deeply for turtles and spend a generous amount of their time trying to protect nesting beaches, rear hatchlings and rehabilitate damaged adults. All of these are very commendable endeavours of which any country would be proud. The difficulty is that some NGOs have totally closed minds when it comes to any form of turtle conservation that differs from their own. They will not hesitate to adopt the high moral ground against anyone, scientist or otherwise, who suggests that sea turtles might be used sustainably. The United States sea-turtle NGOs are numerous and exceedingly enthusiastic and the first world conference attracted a huge share of them.

The world's first and, to all appearances, most promising sea-turtle farm was called Mariculture Ltd. Situated in the Cayman Islands, the farm had been established about ten years previously and had suffered badly when the United States scientific and conservation lobby turned against it. The debate with US scientists paled into insignificance in comparison with the attack by one of the NGOs. They insisted that the farm was, among other things, a blot on society, had no redeeming features, was a threat to the survival of sea turtles and, triumphantly, was suspected of laundering illegal turtle products through its abattoir. The hall was stunned into silence at this atrocious final allegation. I would never have expected this from any quarter, especially as the farm was well known to many of us and was supported by a sizeable group of turtle biologists (see Chapter 18).

To the surprise of many, and to the horror of the president of the conference, the British ambassador requested the floor. On being recognised, the ambassador stated, with typical British understatement, that the remarks made by the representatives of the NGO were 'deeply distressing to the government of the United Kingdom, under whose jurisdiction the Cayman Islands fall, and which represents many of the significant supporters and scientists involved in the turtle farm'. He then stated very quietly that, unless a written and public apology was delivered by 3 p.m., 'Her Majesty's government will institute legal proceedings against the NGO'. This was very new in the experience of many and the fact that it was happening in the US State Department apparently caused a major tremble through the diplomatic world. The world descended on the NGO, which, of course, had no proof at all of its allegations. By 3 p.m. apologies were flying in all directions – from the NGO, from the president of the conference and finally from the chairman of the session. I stole a look at the ambassador and he had the air of a very diplomatic cat that had recently eaten the cream.

This minor setback, unfortunately, did not dampen the fury of the NGOs against the farm, even though in formal conservation circles, and even large-scale NGOs such as the IUCN, there was a growing understanding and appreciation of the value of sustainable use in wildlife conservation. A broad spectrum of species, such as crocodiles, vicuña and many large mammals, have benefited from being ranched and farmed, but, alas, other than killing by indigenous people in their home territories, any suggestion of using sea turtles for ranching and farming has achieved little but raise temperatures in conference halls.

As a strong supporter of sustainable use, I believe that the Cayman venture was not well treated and will devote some time to explaining my reasons for believing so.

CHAPTER 18

# Sustainable Use and the Great Divide

The whole concept of sea turtles making a sustainable contribution to the well-being of humankind and, as a result, being viewed as an asset worth nurturing, was central to my interest in them. My two conservation homes, the Natal Parks Board and the Oceanographic Research Institute, both had sustainable use as a core principle of their policies. The IUCN has long valued the policy as a positive contribution to the conservation of species and has a specialist group devoted to the field. Despite numerous cries, especially from NGOs, that the whole concept is immoral and a crime against nature, this is simply not true. With the growing human population, I believe that this is a policy more likely to save animals and plants from extinction than well-intended and 'moral' appeals to humankind's altruism. A hungry family tends to have a fairly short-term view of natural resources no matter how endangered they may be. If the family can be fed from an available resource, it will be used.

Nature conservation should never have become a question of defending protected areas with arms. We should have been a great deal more intelligent and given people in need benefits from as many natural resources as possible. Alas, this error by many past and current conservation bodies, both formal and informal, has not resulted in quite as many benefits for wildlife as were, shall we say, naively envisioned. There is no doubt that the development of nature conservation and protected areas was a great and altruistic step forward by humanity, but one must understand that it was a very top-down process.

Sea turtles have been subjected to horrendous overexploitation in many parts of the globe. Without active conservation measures, even more populations would have disappeared than have done so. A balanced view from early on should have had conservationists attempting every management strategy to restore the notable value – aesthetic, economic and nutritional – of these wonderful reptiles. It is a great pity that the most influential elements of the conservation world chose the total-protection and animal-rights philosophies that have dominated worldwide turtle conservation since 1969. Has the process had positive benefits for sea turtles? Yes, it certainly has and I have been privileged to share in those successes. Had a broader vision been practised, would the observed successes not have been achieved, and perhaps others as well?

In 1969 many of the rather small band of practising sea-turtle biologists were very much in favour of sustainable-use policies. In South Africa our original policy documents concerning the conservation of turtles in Maputaland stated that when the populations of turtles had recovered to a safe level (a regular nesting population exceeding 500 females per season, in the case of loggerhead turtles), consideration would be given to permitting the harvesting of turtles and their products.

Let me hasten to add two things. Firstly, the local people gave up the consumption of turtle products almost entirely. Secondly, a reason for this may be that tourism, including the local community-managed walking tours, gave the turtles real and meaningful economic value. In addition, I believe that the outstanding participation of so many local amaThonga in the turtle conservation programme over the past 50 years has engendered an affection and pride in the turtle visitors that certainly was not there when the Natal Parks Board started the programme.

In Surinam, Dr Joop Schulz, of the Surinam Forest Service, was already committed to sustainable use because the people of Surinam had collected eggs and killed nesting females for hundreds of years.

What Joop envisaged was a careful balance between permitted egg harvesting and the rearing of turtles from hatchlings to market size in order to reduce the pressure on wild nesting females. His work received support from a number of us in the newly established Marine Turtle Specialist Group (MTSG). At the first meeting, in 1969, there was a wonderful air of mutual respect and encouragement. By the second meeting of the MTSG, in 1971, there had been a significant change in attitudes, however. The catalyst for this change was the Grand Cayman turtle farm, Mariculture Ltd.

It was sad that the problems leading to the division did not appear, at first, to be a matter of principle – whether one should allow exploitation of sea turtles or militate against every form of exploitation. It appeared to be about a clash of personalities into which toxic brew was mixed some over-the-top advertising by Mariculture Ltd. There was no doubt that some of the farm's claims were fanciful, and whether their claim that their products would kill the market for illegal products was a moot point. However, there was nothing that could not be corrected. After all, it is not unusual for advertising agencies to make claims that are beyond belief.

The problem, I was informed, lay in the attitude of one of the Mariculture Ltd executives, who, on being approached by US conservation enthusiasts to ask them to cool down the advertising material, told them to take a hike. The investors had sunk a lot of money into the venture and Mariculture Ltd would advertise as it saw fit. The meeting ended badly and spurred some critical people into turning resolutely against the farm and managed to persuade others, hitherto with open minds, to do the same.

By 1971 this altercation had not yet reached boiling point. The second meeting of the MTSG was noteworthy in that the IUCN recognised that the matter of Mariculture Ltd had to be addressed. They invited Mark Fisher (a director and major shareholder of the farm) to present an overview of the farm's activities and prospects and, as a representative of the consumer market, invited John Lusty, of John Lusty Ltd, London, to present the view of the soup producers. It turned out to be quite an enlightening meeting. If the rather innocent and naive understanding of turtle biology among biologists at the first meeting of the MTSG had been obvious, it became even more obvious at the second meeting that the approach of the farm and soup businesses was equally naive. What was required was a better attempt to achieve a meeting of minds. What actually happened was the beginning of the great divide that remains in place to this day.

Feelings were strongly expressed at the meeting, as there were some terrible reports of overexploitation. One report suggested that well over a million sea turtles a year were being killed for meat, soup, leather, tortoiseshell and bones. Strangely enough, not a single word was said about the value of turtles as a tourist attraction. As the meeting progressed, a clear division developed, which ultimately split the members into pro-farming and anti-farming camps. The pro-farming camp was composed of biologists mainly from Third World countries: Dr Bob Bustard, Australia; Dr Joop Schulz, Surinam; Dr E Balasingham, Malayasia; Dr Stan da Silva, Sabah; Dr Leo Brongersma, Holland; and myself. The United States members, with the exception of Dr John Hendrickson of the University of Arizona, all moved firmly into the anti-farming camp.

By 1974 the IUCN found itself under increasing pressure to speak out against Mariculture Ltd. In response to this, the Species Survival Commission decided to hold an inquiry into the farming endeavour and invited a number of us to join the task force that would manage the hearings in Miami, Florida. The hearings would be followed by a visit to Grand Cayman and an in-depth visit to the farm. I accepted with alacrity and the Natal Parks Board, which I had just rejoined as a senior professional officer, approved my visit.

The inquiry should not have been held in Miami. The constant string of turtle enthusiasts arriving to abuse the farm became a trial. Some of the anti-farming scientists exceeded the bounds of reason in claiming that the farm should never receive endorsement from the IUCN and that the IUCN should condemn the entire endeavour. One US scientist (who later became an editor of a prestigious and extremely well-respected journal) was so rabid that he even claimed that the farm used too many resources, such as diesel to keep its pumps running

– this from an American whose country had just spent in the region of US$345,000,000 on sprucing up the site for the Winter Olympics in New England. The fact that the farm employed around 80 full-time staff, many of whom were native Caymanians, and the farm had become *the* major tourist attraction of the islands, was considered of no value.

Fortunately, there were a number of very courageous and articulate US scientists, such as Dr John Hendrickson, who were actively engaged in research in collaboration with Mariculture Ltd. They expressed the view that there were great possibilities for the farm and that it should be given a fair chance to succeed or fail.

This accorded with my view entirely and remains my firm conviction.

The visit to Grand Cayman was splendid and very informative. Two of the United Kingdom's best-known scientists, deeply involved in research on the farm, were present. Sir Alan Parkes and Prof. EP Amoroso were embryologists/physiologists of world standing and their addresses on their work were fascinating. There were, at that stage, over 40,000 green turtles in the farm and the resident turtle biologists, Drs Jim and Fern Woods, were making great progress in discovering the pitfalls of turtle farming. All manner of problems that had arisen in the hatcheries, including the transport of eggs, hitherto unknown diseases (fungal infections in particular), as well as many other ailments not yet recorded, were described and treatments developed to deal with them. To me this was an exciting enterprise, full of risk, of course, but with great promise. The IUCN group that visited the farm left with, I thought, a fairly positive impression, and were certainly convinced that the venture should not be condemned outright.

The generally positive feeling of the visiting group never really made it into print and the draft report emanating from the IUCN was less than enthusiastic and, in any case, had been overtaken by events. The US scientists lobbied frantically in the US to close all markets, effectively killing the most lucrative financial lifeline for the farm. Americans, at one stage, ate a lot of turtle meat and consumed a lot of turtle oil, especially in the cosmetics industry. An attempt to open a wider market in California was attacked with such vigour that the idea was stillborn. In time, the US even managed to prevent the shipment of farmed products through US ports, thus adding additional transport costs to the farm. It was one of the most virulent campaigns against sustainable use in the history of sea-turtle conservation.

The reasons put forward by the anti-farming lobby were:
- The use of farmed products would expand the market for turtle products, which, it was claimed, could not be filled by the farm and would therefore lead to a runaway increase in turtle poaching.
  *Comment:* the possibility that, if Mariculture Ltd proved a success, other farms might be developed around the world (many countries have healthy green-turtle populations) to supply a growing demand was not given any consideration.
- The farm would produce an expensive set of products essentially for only the upper-class market.
  *Comment:* this from the land of the free and free enterprise!
- The farm purchased eggs from Surinam and Ascension Island, which were seen to represent different populations. The possibility that accidental escapes from the farm would affect the gene pool of local green-turtle populations was suggested. To make it worse, the farm was releasing 'chicken turtles' roughly a year old into the waters of the Cayman Islands, thus adding to this supposed risk.
  *Comment:* it should be pointed out that, despite having been home to one of the largest concentrations of nesting green turtles ever recorded, the Cayman Islands had not one wild nesting turtle left at this stage. I was appalled that these conservationists were committed to seeing no green turtles ever returning to the islands, rather than have a slightly mixed gene pool. Whatever differences may have resulted would probably have been beyond determining in any case, as the distribution of the green turtle is pantropical and they are all included in the same scientific species, *Chelonia mydas*, right around the world. (There have been some attempts to give subspecific status to some colour morphs in one or two eastern Pacific populations, but even Archie Carr was not convinced, especially

when one enthusiastic taxonomist suggested that one population be called *Chelonia mydas carrinegra*. Archie was not flattered and said it would be referred to as the 'Black-hearted Carr's turtle'.)

- The production of high-quality tortoiseshell from the farmed green turtles would further stimulate demand for wild tortoiseshell normally obtained from the hawksbill turtle.
  *Comment:* to our surprise, we were shown tortoiseshell from the farmed green turtles, which, as a result of their high-protein feed, was very much thicker than that found in wild green turtles. Wild greens have very thin tortoiseshell that is of no commercial value. In fact, tortoiseshell from the farm was of a better quality than that obtained from wild hawksbill-turtle populations, the normal source of this valuable product. In addition, because the turtles were in tanks, the usual reef-scarring of the tortoiseshell plates, so common in wild populations of hawksbills, was not present, so there would be less wastage in the working of the tortoiseshell. The possibility that a supply of tortoiseshell products from the farm might help reduce pressure on wild populations of hawksbills was given scant consideration.

The farm was never officially endorsed by the IUCN and what positive comments were made were almost apologetic. The draft report received severe criticism from the pro-farming members and, to my knowledge, was never officially published.

In order not to be accused of being committed Luddites, the anti-farming US scientists came up with more and more bizarre hurdles that the farm would have to clear before it could be recognised in terms of the CITES rules that had recently emerged. The first was that unless the farm could produce an F1 generation of turtles (green turtles in captivity, raised for sale, born from parent turtles resident in the farm), it would not be recognised. When the farm announced that it already had some F1 turtle hatchlings, the anti-farming lobby quickly campaigned to change the rule, demanding that the farm have F2-generation turtles for processing before recognition could be considered.

This meant that the farm would have to raise the F1 generation to mature nesting size (perhaps 20 years) and would then have to successfully mate them, producing enough F2 generation animals for processing and sale to maintain the farm as a viable concern. This was a period of possibly 23 years before recognition.

In the face of this orchestrated and malicious onslaught, the original investors and pioneers declared bankruptcy in 1975 and the farm was partly taken over by the government and a family of German investors. Unfortunately, the anti-farming campaign never lost momentum and continues to this day. Eventually, with the withdrawal of the German investors, Drs Judith and Heinz Mittag of Dusseldorf, grievously hurt by the constant attacks on their integrity, the farm ceased to be a vibrant endeavour and reverted to being far more of a tourist attraction maintained by the local government. Products from the farm still made their way to the British soup market, but the fire was gone and the anti-farming group ultimately triumphed.

Incidentally, the side effects of a hurricane virtually emptied the farm of turtles a few years ago and it has been moved, along with a nature park, into a less vulnerable part of the island. The centre currently enjoys a patronage of 500,000 visitors a year.

However, the tone had been set by the anti-farming lobby and the tools were now in place in CITES for opponents of turtle farming or ranching to frustrate any other enterprise involving the sustainable use of sea turtles, and they did.

One is reminded of the theme of Mark Anthony's speech in Shakespeare's *Julius Caesar*: 'The evil that men do lives after them; the good is oft interred with their bones'. At the very least, those who shared the vision of a turtle farm must have empathised with Mark Anthony's words.

CHAPTER 19

# Enter the Mascarenes

While all the drama aimed at destroying the Mariculture Ltd sea-turtle farm was being enacted, I was trying to encourage other countries to engage in turtle farming.[1] Successful turtle farms would provide jobs, promote good conservation techniques and save the green turtle from overexploitation in the wild. Of course, I had no great claim to being an original thinker in this regard. Archie Carr had already stated that, 'A technology of green-turtle husbandry will have to be developed. Once that is worked out, it will be a double blessing: people will be fed and a species will be saved.'[2]

There were a number of reasons why the south-western Indian Ocean was a suitable region to attempt such ventures. Firstly, there were many green-turtle nesting colonies available from which to draw stock, either as eggs or as hatchlings. Secondly, many countries in the region have year-round warm sea temperatures, which is essential for year-round turtle growth. Thirdly, job creation was, and remains, an imperative because most countries in the region are Third World.

By 1971 I managed to visit and describe nearly all of the major green-turtle nesting sites in the region, including some that had never been described and others whose importance had not been realised, such as Europa Island in the Mozambique Channel. My esteemed colleague Jack Frazier (later Dr Jack Frazier), had done excellent work in Aldabra and had been the first to alert the turtle world to the existence of large green-turtle nesting areas in the Comoros. Between us, we announced to the world that the region was very rich in sea turtles, especially green turtles. The presence of some visionary souls who would be prepared to invest time and money into a long-term development was needed. Developing a turtle farm is not an inexpensive affair, but, by great good fortune, such people existed in both Reunion and Mauritius.

The Mauritius potential, in particular, was phenomenal. Firstly, I had the good fortune to be befriended by Mr Keith Cox, an FAO development expert stationed in Curepipe.[3] His task was to promote better fisheries husbandry and act as FAO's man on the spot. Keith had quite a hard time introducing First-World standards to the local Creole community, but he was totally dedicated to his task and embraced my suggestion that a turtle farm was a possibility for Mauritius.

Some 400 kilometres north of Mauritius lie the St Brandon islands, a massive coral-reef complex about 40 kilometres long with about 26 sandy islands on which green turtles were reputed to nest. Mauritius exploited the green turtles there for many years and even after my first visit to Mauritius in 1971, turtle products from the islands were finding their way to Port Louis in modest quantities. The Mauritius Fishing Development Company maintained accurate records of imports of green turtles from St Brandon, but under moderate pressure from the government all such traffic was stopped in 1973. Following my research visit, it was confirmed that many of the islands had nesting green turtles, so a source of farm stock, in the form of eggs or hatchlings, was conveniently available.

Even more promising was a site that would cost virtually nothing to develop as a farm. Near Mahébourg, on the east coast of Mauritius, was a large and very expensive installation, which had apparently been a spectacular failure as a fish farm. It was an impressive complex with a broad spectrum of concrete pens, an efficient network of pipelines serving the tanks with fresh seawater and, most important of all, an entire bay, called a *barachois* by the locals. The *barachois* was walled in from the sea and enclosed nearly 200 hectares of natural reef and basin. It had two openings to the sea, barred with metal grills, providing free tidal interchange of seawater and incredible quantities of naturally growing algaes, a prime food resource for green turtles. Keith was using the complex as his base for running trials on oyster culture, but the complex could not have been better suited to turtle farming. Keith and I were ecstatic and I wrote an extensive set of recommendations to the Mauritius government recommending the establishment of a turtle farm.[4]

The proposal was received with some enthusiasm by the Mauritius Department of Fisheries. Keith was given permission to start in-depth planning on site and even his FAO seniors were supportive of the scheme. More impetus was given to the proposal when Mr Paul Cousseran, the governor of Reunion, agreed to make stock available for the farm from Les Îles Éparses. This was a good example of inter-Mascarene cooperation, which was much appreciated.

By 1973 Mahébourg received its first shipment of hatchlings from St Brandon, which took everyone by surprise as the upgrading of the farm had not begun. The young Peace Corps volunteer who was tasked with the day-to-day supervision of the farm told me that their losses were quite high, but eventually some healthy 20-centimetre turtles were released into the *barachois*. And, that, I regret to say, was as far as it went. Keith was transferred out of Mauritius by FAO and the whole scheme collapsed. I do not believe that Keith's transfer had anything to do with the possibility of ranching sea turtles, but the farm had lost its champion and Mauritius had missed a great opportunity. One of the reasons given for its establishment was the farm's potential as an attraction for the tourist industry beginning to develop in Mauritius. Since then, tourism has become a massive earner of foreign exchange for the small island and the farm attraction would have been as great an asset as similar farms in Grand Cayman and Reunion.

Champions are necessary for any project to succeed and happily, in Reunion, I found three. The first was the dynamic director of the Meteorological Services, Mr Marcel Malick. Short, red-haired, fiery and absolutely dedicated to the well-being of Les Îles Éparses and their turtles, no one could have asked for a better friend and colleague. He took to the thought of a turtle farm in Reunion with almost reckless abandon. Not only did he have the massive resource of thousands of nesting greens, laying millions of eggs per year, at his beck and call, but, more importantly, he had two other champions: Mr Paul Cousseran, the governor of Reunion, and Dr Christian Jouanin, an international expert on sea birds and a scientific advisor to Reunion. Mr Cousseran was so fond of sea turtles, following a visit to the islands, that he even had hatchlings in his swimming pool. They saw to it that several students, working in civil positions instead of doing military service, were despatched to the islands to follow up on my visits and one young man, Alain Lebeau, was brought to Reunion to become, effectively, the first curator of the turtle-farm project. Mr Cousseran arranged an alliance between the Department of Fisheries and the Meteorological Services and, near La Saline les Bains on the west coast, they found some old concrete tanks. These were patched up, pumps were fitted and the area was readied for stock. Marcel put some departmental money into the project and, as he put it, 'My contribution is to bring in the first bunch of hatchlings and tough luck on the frigate birds!'

The stage was set for a greatly enhanced collaboration between South Africa, Mauritius and Reunion. Both Christian and Mr Cousseran were enthusiastic about establishing a scientific committee including all three countries, if not more, and Mr Cousseran made an official visit to South Africa to discuss, among other things, this possibility. It was a useful visit, but he viewed South African politics with great concern and he drew me apart at a function in Durban and said that South Africa must change, '*Sinon, vous serez mangées*!' (If not, you will be eaten!) Of course, he was correct about the need for change, which took another 20 years, but happily he was wrong about the process.

Regrettably, Mr Cousseran was transferred back to France shortly afterwards, which put paid to the scientific committee for about 20 years. The farm, however, under the enthusiastic patronage of Marcel and the Meteorological Services, grew and prospered.

A new site for a bigger and better complex was chosen at St Leu and the farm CORAIL (The Reunion Company for Aquaculture and Littoral Industries) was established in 1977 and eventually grew into a complex of significance. Within three years it was in production, rearing green-turtle hatchlings brought from Europa. By 1984 some of the adult females were nesting and laying eggs on the beach bordering one of the larger ponds. As with Grand Cayman, the farm used high-protein pellets, imported from France, to feed the turtles, which grew well, reaching optimum slaughter size at three years. The farm proved a success and sold a wide variety of meat products from soup, steak, dried meat and mince to heart, liver and pâté. They also sold leather and the most beautiful products derived from the amazingly high-quality tortoiseshell. A talented group of tortoiseshell artisans established businesses there as a result and the farm became one of the top tourist attractions of Reunion.

Many of these products were exported to France, which helped to make the farm sustainable. The farm was widely regarded as a success as it paid its way, employed a large number of local people and the quality of its products improved every year. The farm also released a number of yearling green turtles into the sea around Reunion each year. The hope was that they would grow up to nest locally and start to recreate the Reunion nesting colony that had once been massive. It is pleasing to report that since 1990, when the first nesting female on Reunion for probably nearly 200 years was recorded, a growing number have been observed nesting each year. Even more encouraging are the regular aerial counts of sea turtles off the west coast. These counts, obtained from microlight aircraft on a monthly basis, are demonstrating substantial increases in the numbers of resident turtles.

Research on Les Îles Éparses was stepped up in order to ensure that any negative impacts could be quickly identified. The farm sponsored and coordinated new tagging programmes to continue my earlier work. Reasonable people, in my modest opinion, would have rejoiced at the farm's success and its conservation endeavours. Regrettably the world is not filled with reasonable people.

As the farm began to prosper, Greenpeace (France) began to agitate against it. They claimed that it violated the CITES rules. The French authorities, charged with the responsibility of dealing with CITES, argued that Reunion was a department of France (what in South Africa would be called a province) and that the movement of products from Reunion to metropolitan France was domestic trade and therefore not subject to CITES regulations. Greenpeace then pressured the European Union, which in turn pressured the French. In 1984 the French government appointed a commission to review the activities of the farm. Five international scientists were appointed to the commission: Dr A. Lebeau, French Polynesia; Dr Nicholas Mrosovsky, University of Toronto; Dr Jack Frazier, University of Arizona; Professor Brygoo, Museum of Natural History, Paris; and myself, assistant director of the Natal Parks Board. Dr Bernard Bonnet, from the University of La Réunion, also joined us. The commission was the responsibility of Dr Arrignon, secretary of state for the sea, Paris. All the commission members could speak French.

The commission spent five days in Reunion and visited Tromelin Island (see Chapter 27) to view the nesting beaches. The entire process of hatchling collection, transportation, rearing, feeding, slaughter and sale was reviewed. The finances of the farm were inspected and both employees and artisans interviewed and, in the latter case, the manufacture of their artistic products was watched with interest and appreciation. Some of the food products were eaten, although I drew the line at the marinated turtle livers, which I, being of British origin, viewed with some suspicion. The steaks, however, were excellent.

When the commission had completed its report there was time to spare, as flights were not all that regular in those days, so our local hosts rewarded us with a helicopter flight over the island. This is truly one of the world's most exciting and unforgettable experiences. Reunion is spectacular. It is a volcanic island barely 35 kilometres in diameter, rising to over 3,000 metres to the peak of the Piton des Neiges, which is surrounded by four massive *cirques* deeply incised into the sides of the island, at the east end of which is an active volcano. On the flight, one passes over these deeply incised valleys, past waterfalls over 300 metres high, through terrifyingly narrow gorges and, after over-flying the volcano, rises to the top of the highest peak, circles it and lands in the most beautiful *cirque*, Cilaos. It was a great end to a successful visit.

A full report endorsing the farm was unanimously supported and the French government appeared satisfied with this conclusion and permitted the farm to operate without interference. Many techniques were developed on the farm, not the least of which, to complement the turtle enterprise and improve the efficient use of feed and pond space, was the successful culture of Mozambique red tilapia. The farm, at one stage, was exporting some 25 tons of tilapia to West Africa annually. In addition, the local veterinarians became skilled in handling turtle problems and the farm became an even greater tourist attraction. Young members of the farm's staff were despatched over several years to the Natal Parks Board programme in Maputaland, where they joined our local teams and learnt from our failures and successes. Gradually, the importance of the farm grew as it spearheaded wider research into turtles and the care of the islands.

Success did not stop Greenpeace and other NGOs from constantly haranguing the French government, however, and in 1998 the sale of turtle products from Reunion to France was stopped. I am sure that the anti-trade NGOs are proud of this. It effectively cut off the high-value end of the market and the farm began to struggle financially. With the removal of high-earning power for the farm, and Reunion, local government support began to waver and the farm began to deteriorate badly. It all appeared to be going the same way as Grand Cayman, but another champion was in the wings.

This was a young Frenchman, Mr Stephane Ciccione, who was managing CORAIL at this crucial time. Stephane is a passionate lover of sea turtles and a firm believer in the integrity of the farm. He was also a great communicator and, using the farm's obvious value as a tourist attraction, persuaded the new government in Reunion to consider rebuilding it as a turtle observatory and the centre of turtle research in the region. The status of the turtle islands as globally important nesting centres was an invaluable tool for him, as was the farm's obvious value as a tourist attraction, and eventually Stephane succeeded in getting a special grant of some €16,000,000. The entire complex was upgraded and rebuilt, and in 2005 was reopened as Kelonia, an institute of which any country would be proud.

Kelonia is a model of its kind with a large display tank containing living representatives of all species of sea turtle, except the leatherback. The displays scattered throughout the complex are modern, global in information and executed with artistic French flair. There is an audio-visual centre with interactive displays (where even I and my colleague Jeff Gaisford, now retired from Ezemvelo KZN Wildlife, give televised talks about our programmes in South Africa) and an auditorium with regular showings of films, many made locally and on Les Îles Éparses. Kelonia has become a centre for exhibitions, photographic displays and, above all, the world-class staff-run environmental education programmes.

School groups are given excellent lectures and tours of the institute, and are taught songs about sea turtles, which they sing when they accompany a turtle being released into the lagoon. This may be a rehabilitated turtle or a tagged turtle fitted with a satellite transponder. Having attended one of these school visits, and having shared in the excitement and enthusiasm of the children, I believe that for this programme alone Kelonia deserves exceptional credit. It is producing a generation of Reunionese who love sea turtles.

But, significantly, Kelonia has stayed faithful to its roots. Although no longer an exporter of commercial sea-turtle products, it has not abandoned the artisans that helped make CORAIL a success. One of the fine exhibits within the complex is a hall devoted to tortoiseshell and the beautiful products that are derived from it. Every day one of the local tortoiseshell artisans mans the exhibit and demonstrates the preparation of tortoiseshell and the manufacture of tortoiseshell jewellery and marquetry. Children visiting the complex are not fed an insipid diet of animal rights and preservationist dogma, but a balanced and reasoned understanding of the cultural, economic, aesthetic and biologically exciting value of turtles. Reunion is privileged to have Kelonia and I am immensely proud to have been associated with it and its predecessors.

Reunion has become one the most important turtle-research centres in the world today. Together with the local staff of IFREMER (the French Research Institute for Exploration of the Sea), staff from Kelonia are deeply involved in turtle research in Madagascar, Mayotte, the Comoros, as well as Les Îles Éparses and Reunion, and are making a massive contribution to the conservation of sea turtles.

The Mascarenes have made a contribution they can be proud of.

1 The Mascarenes are the three islands of Reunion, Mauritius and Rodriguez.
2 Carr, A. 1967. *So Excellent a Fishe*. Garden City, New York: The Natural History Press.
3 United Nations Food and Agriculture Organisation based in Rome, Italy.
4 Hughes, GR 1972. 'The Proposed Mahebourg Turtle Farm'. Durban: Oceanographic Research Institute.

CHAPTER 20

# Mozambique: The First Attempts

Perhaps carrying out research in the midst of a civil war is not the most sensible thing to do. Throughout life one is faced with choices and I can honestly say that the gods have stayed firmly on my side. The choice I made to go ahead with the Mozambique survey was as sane and lucky as my decision, a few years later, not to go ahead with an Angolan survey. Mozambique is a beautiful country and my experience with all manner of people there suggested that they were pretty friendly and helpful. Certainly, after the enforced racism so prevalent in South Africa, the social atmosphere in the countryside was delightful. On top of that, the green wines were excellent and the prawns superb.

With 3,000 kilometres of coastline to survey, I was a little apprehensive when I started off in late 1969. I could hardly speak a word of Portuguese, my contacts were minimal and information on the road systems and access to coastal areas were, as a result, fairly flimsy. I had briefly visited Mozambique's southern beaches in pursuit of turtle information twice before. In 1967 I travelled up with a Natal Parks Board party of board members and staff that was invited by the Mozambique Department of Veterinary Services. My presence was at the personal request of Mr José Tello, the warden of the Maputo Elephant Reserve, south of Delagoa Bay, and I was welcomed by the Natal Parks Board staff because they now had a baggage driver. José proved a brilliant host and I had a good look at the coastline around

Ponta Milibangalala, south of Inhaca Island, and the beaches near Lake Piti. The coastline is almost identical to the beaches on the South African side of the border and we clearly shared a meta-population of loggerheads and leatherbacks.

Eager to find out how far north this population extended, I was rash enough, in 1968, to accept an offer from Paul Dutton, the warden of Ndumu Game Reserve, and the brand new owner of a pilot's licence and a Piper Super Cub (famous today as the *Spirit of the Wilderness*) to fly the coast. As the area was somewhat short of landing strips, Paul casually stated that he would use the beach and, true to his word, at low tide on the appointed day, he put the plane down on the beach near Bhanga Nek. The plane looked brilliant swooping down onto the beach and, without adequate insurance, I leapt in and we did our first sortie down to Cape Vidal and back. Paul was even more gung-ho than I expected and on seeing some leatherback nesting tracks he decided to put us down on the beach so that we could check whether the female had laid eggs. There was some justification for this because the leatherback population was extremely low, averaging 20 females per year over the first ten years and Paul had never seen a track close up. It was not the best of decisions.

Landing on a beach requires a fairly intimate knowledge of beach dynamics when viewed from the air, a skill that neither of us possessed. Landing at a dead low tide when the beach is flatter than a normal airfield is one thing; landing half a tide later when the amount of beach available is much reduced and consists of a sharply rising slope, gently moulded by the incoming water into steep cusps at short intervals, is quite another. If Paul wanted a test of the value of a robust little aircraft that flew, without falling out of the sky, at the incredibly slow speed of 64 kilometres per hour, this was it. With eager anticipation, we hit the beach and all went well until the first cusp, at which contact we flew into the air, very nearly breaking our necks, landed again, hit the next cusp and repeated the leap into the air. By now the two of us were hanging onto anything that projected and after about six repeats the plane stopped. It was a sobering lesson in beach landing, used in future with greater circumspection. As a matter of interest the leatherback had nested.

## Mozambique: The First Attempts

The next day Paul and I set off in the *Spirit* for Lourenço Marques (now Maputo). It was a lovely day and we were disappointed to count, at the height of the nesting season, very few nesting tracks of either species, all the way to Inhaca Island. We recorded a mere fraction of what was nesting south of the border in Maputaland. This was not encouraging, as we had heard that the killing of nesting females, especially loggerheads, was very common along this stretch of the Mozambique coast.

After a rather splendid night in LM with some of Paul's friends, we flew up the coast to Xai-Xai and then turned back to South Africa to land at Komatipoort. We hardly saw a nesting track on the beaches.

Paul, you must remember, was a very new pilot, so when we landed in Komatipoort the experience was very similar to our landing on the cusped beach. As we entered the terminal, a bush pilot of ancient vintage, judging by his bushy handlebar moustache and nonchalant air, was leaning on the bar and asked which one of us was the pilot. Paul, with an equally nonchalant air, acknowledged that it was he. The old pilot then took out one of the very large R1 coins that weighed our pockets down in those days and, sliding it across the bar to Paul, said, 'Then this is a contribution to your next landing lesson'.

The following year, in late 1969, having only recently purchased the Panda Wagon, I managed some rushed organisation before heading north, armed with tagging and measuring equipment and enthusiasm. You can be sure that I was underprepared. When I reached Lourenço Marques, I contacted José Tello, who immediately offered to accompany me up the coast as far as the Save River. Beyond this, he informed me, I could not go, as the Save bridge was not yet built and the river was in flood and impossible to cross. Armed with a letter of blessing from the authorities, which included an appeal to help the researcher, we set off north. Wherever possible, we stopped at each coastal site and walked the beach. We were disappointed at the lack of evidence of nesting and even more disappointed when most local fishermen said they ate turtles when available.

By the time we reached Inhassoro, on the mainland opposite the Paradise Islands, we were of the opinion that turtles were a rarity along the coast. We were fortunate indeed in finding two ex-Rhodesians, Tom and Margaret Keightley, who had settled in Inhassoro and made a living out of taxidermy by preparing small fish and other marine life as souvenirs. They were very well informed on anything coastal and told us, to our great relief, that there were many turtles in the region. The sea between the mainland and the islands was the first prime green-turtle feeding habitat that I had ever come across. Shallow, warm water over massive beds of sea grasses teeming with fish is an ideal feeding habitat for green turtles and, as I learned later, dugong.

This environment contrasted greatly with the reef-and-algae habitat commonly associated with green turtles in Maputaland and all the way down to Cape Agulhas. Green turtles are far from uncommon in South African waters, but they do not nest along our shores. Year-round warm temperatures are preferred for nesting greens and here, for the first time, I saw the richness of such feeding grounds. There were plenty of green turtles around, but there was also a surfeit of proof that many of them were being killed.

Over many decades, a Chinese community had established itself in Inhassoro and had developed a simple but successful land-based trawl system of fishing. Each morning, at various points along this extensive and vehicle-friendly beach, small boats would be launched carrying huge bundles of nets. One end of each net was attached to a tractor-driven winch above the high-tide mark and as the boat moved offshore the nets were played out for hundreds of metres. About 600 metres out to sea, the boats would turn in one direction or another, continuously feeding out the net and, gradually following a circular route, would make their way back to the beach. The other end of the net was then attached to another winch driven by a second tractor.

After a brief interval, the tractors were started up and the trawl net was gradually dragged across the sea bed, becoming narrower and narrower until it hit the beach. The entire net was then drawn out of the sea and onto the beach where the fishermen immediately began sorting the contents. I need hardly add that this form of fishing collected everything on or near the sea bed. Not only did the catch contain fish (some useful and others not), but also rock lobsters, prawns, crabs,

## Mozambique: The First Attempts

not uncommonly turtles, huge quantities of *Cymodocea* (an undersea angiosperm and favoured food of the green turtle) and general debris. The Keightleys were into the catch along with the others, having permission to take all the colourful butterfly and porcupine fish, which were of no value to the Chinese owners.

The presence of the *Cymodocea* explained why the green turtles were caught and, although none was caught during the trawls that I witnessed, there was ample evidence at every fisherman's complex along the beach that, over time, many had been. All turtles were killed and the meat either eaten or sold. Large green turtles were dragged on their backs to the owner's home and one could see where the sand had literally sand-papered the scales off the carapaces of the larger turtles. Smaller turtles were simply thrown onto the trailers along with the other catch and the nets. José and I measured a large number of carapaces and plastrons lying in the bush near the cottages, and it was the first decent sample of turtle material that I obtained in Mozambique. A number of loggerhead carapaces were found, but the Chinese did not report finding any tags from South Africa, despite my explaining the value of the tag and the reward that we paid for each one returned.

No turtles nested along this part of the coast, but we were informed that turtles nested on the Paradise Islands. Years later, Paul Dutton spent some time on the main island of Vilanculos establishing a turtle-monitoring programme under the auspices of the South African Nature Foundation. He found loggerheads and the occasional greens nesting there. More recently, as the political situation in Mozambique has settled and tourism has developed in the region, it has been established that loggerhead turtles nest in modest numbers on the mainland just south of the Paradise Islands along the beaches of the San Sebastian peninsula.

Although we were not aware of the fact at the time, José and I were at the extreme northern end of the nesting grounds of the south-east African population of loggerhead and leatherback turtles.

Our twenty-first-century satellite tagging has demonstrated quite clearly that in the deeper waters offshore, and around the islands, many of 'our' nesting loggerheads have feeding territories, so it is hardly surprising that loggerheads were caught in those nets. This does not happen anymore because after 1975 the Chinese departed the region and the land-based trawl fishery ceased to exist. This was good news for the turtles, and the occasional dugong, that perished from time to time. Many of the residents in the area, such as the hospitable and invariably kind Keightleys, who moved to South Africa, also disappeared.

As a postscript with respect to this region, which I visited many times in the following few years, I was particularly pleased when Dr Ken Tinley, the resident ecologist at Gorongosa National Park, managed to persuade the Mozambique government to declare the Paradise Islands Marine National Park, using, as part of his motivation, the presence of turtles in the area. This park has survived the political changes in Mozambique and is once again thriving as an area of exceptional beauty.

Of course, the Chinese have now gone and they have taken with them their unique influence and practices. Not having been there recently, I wonder whether the new postmasters have changed the Inhassoro post office. It was rather special, being fitted into an incredibly small space under an external flight of concrete steps leading to the second floor of the building. There was barely room for the postmaster behind the counter and behind him, in full view, were the post boxes obviously belonging to the Chinese, one of which, I recall, had the name 'Low Fuck Jim'.

José and I reached as far north as we could drive before returning to Maputo. I left him with many thanks to his beautiful and gentle wife Grace for lending me her husband and hosting me overnight. José was actually quite a forbidding character. He was a hunter of great fame in the area and disappeared for months into the hunting *coutadas* of Mozambique. He was a passionate conservationist and was employed by the Servicos Veterinario, which promptly whipped him back to Gorongosa after I departed. Judging from the reaction of his gorgeous little daughter to his gruff attempts at affection, I do not think that she had spent much time with this rough and ready character with the penetrating black eyes. José was a proud and committed *Mozambicana* and for many years, long after independence, served his country as a hunter, manager and biologist.

*Mozambique: The First Attempts*

As far as the civil war was concerned, I found that the further away from Lourenço Marques I travelled, the less concerned about it the local population appeared to be. In the Inhassoro area, the war was almost completely absent and the only reference to it that I heard was that someone had been attacked by a local tribesman with a bow and arrow.

Now, however, my travels were going to take me north into regions where the war was not simply a source of rumours, but a clear and present danger.

CHAPTER 21

# 'A gross abuse of public funds!'

There are detours and there are detours, but I never guessed at the one that would lead me back into northern Mozambique in December 1969. I re-entered South Africa from Lourenço Marques after establishing contact with Dr Tony de Freitas at the Missao de Estudos Bioceanologicos e de Pescas de Mocambique (The Institute for the Study of Bioceanology and Fishes of Mozambique) and Roger Oxley-Oxland, who had purchased a fleet of prawn trawlers operating out of Antonio Enes (now Angoche) in the far north. I motored through the Transvaal, across Southern Rhodesia and back into Mozambique to cross the Zambezi at Tete. If ever there is a site epitomising white man's grave country, the unbelievably hot and humid Tete is it. The very walls on the buildings were covered with verdigris and looked as if they had not seen paint since the Portuguese first settled the area some 400 years previously. I had my first experience of travelling in convoy on this route, as there had been attacks by Frelimo guerrillas on individual vehicles on the lower Tete road. The powers that be believed in safety in numbers and about 20 vehicles made up the convoy.

Impressed by the size of the Zambezi River as we crossed it, I drove on to Malawi, arriving in Blantyre where I was befriended by the US Peace Corps who looked after me while I planned my northward move to Tanzania. I was armed with a sheaf of impressive testimonial letters espousing my research programme from the IUCN, the WWF, the

## 'A gross abuse of public funds!'

MTSG, the Southern African Wildlife Foundation and others, so I was confident that I would have little trouble entering the country. I was wrong again.

I spent Christmas Day in Blantyre and, quite fortuitously, overheard the Christmas message by President Hastings Banda. I confess to being surprised at its content, having been used to more bland fare from Queen Elizabeth, or more religious rants from the South African presidents. Dr Banda was obviously feeling a little paranoid because his message focused on the possibility that agitators were entering Malawi from the north. He then exhorted his loyal populace, if and when they encountered strangers in their village, not to phone him and tell him, but to 'Take your pangas, kill the strangers and *then* phone Hastings Banda and tell him there are strangers in the village!'

This paranoia was fairly widespread. En route to the north, I was stopped by a large troop of AK-47-wielding Young Pioneers who, for some unknown reason, regarded me as a stranger. As their average age appeared to be 14, I was more than mildly petrified. Putting on a brave face, I leapt out, grinning inanely, and told them who I was and why I was there. Unmoved by my explanation, they stripped my vehicle, scattered my belongings and opened everything. It is difficult to cling to one's belief in fair play and non-racism under such circumstances, as I felt their attitude was unnecessarily belligerent and unsporting, especially when they found my speargun lashed to the inner roof of the Land Rover. The discovery resulted in my facing the business end of 35 AK-47s while the troop retreated five metres to get a clear shot. Their leader had me remove the speargun with great care (which I did) and demonstrate how it was used. You may smile, but I actually told them that loading a speargun out of water is an offense. They were unimpressed. When I put a spear through a handy cardboard box, it brought out the children in the lot of them and I spent the next half hour (willingly) giving them all a turn at discharging the speargun. They were clearly warming to me, and I was even asked to hold someone's AK-47. Only the interests of sensitive readers forbid me from telling you what passed through my mind at the time.

One of my best friends, and a fellow graduate from the University of

Natal, Anthony Hall-Martin, had recently accepted a post at Kazungu Game Reserve, virtually next to the road to Tanzania in mid-Malawi. Anthony spent a few days trying to calm me down by taking me far too close to elephants for comfort. Little did I know that his outlandish fondness for elephants would, via a master's degree on giraffe behaviour, eventually lead him to a doctorate and considerable fame. Anthony had lost a game guard that week. Prior to my arrival, the guard had been speared in the leg by Zambian poachers and taken to the nearest hospital for treatment. The resident physician asked Anthony to return in a week's time, which he did, taking me with him. To Anthony's horror the physician told him his guard was dead, despite the fact that it had been only a leg wound. When Anthony protested, the medic replied that he should not be concerned 'as many patients die here'.

After that confidence-inspiring contact with the Malawian hospital system, I drove north, eventually reaching, via Zambia, the Tanzanian border post of Tunduma. I had spent a few nights there in 1960 while hitchhiking back from Europe. I recalled it well because the immigration hall was neat, tidy and well maintained and housed. At that time, Northern Rhodesian customs officials were faced across the hall by expatriate English customs officials serving Tanganyika. They disliked one another intensely. After I had passed through immigration from the north, darkness fell. No other vehicles passed through and I wanted to go back into Tanganyika to the only modest hostelry for about 200 kilometres in either direction. The Tanganyikan immigration people refused to let me back in (even then I was obviously an object of suspicion) and one of the Rhodesian officials simply walked across the hall and threatened to beat up the Englishman, who had a sudden change of heart.

This time I was less successful. Firstly, the same customs and immigration hall belonged totally to the Tanzanians and it appeared that during the past decade a minor war had broken out inside the hall. It was an absolute shambles, with broken furniture piled up against the walls, and it had clearly received no maintenance since my last visit. The customs officials were not exactly welcoming and, while I was welcome to go in for free, they demanded a cash-on-the-spot payment

## 'A gross abuse of public funds!'

of £400 to allow my vehicle through. That was not the sort of cash I carried with me and, in any case, I was not convinced that the *Carnet de Passage* was going to carry me very far. So I haggled.

While engaged in this futile endeavour, I was surprised to see President Julius Nyerere bounce through the door, greeting everyone, except me, with incredible friendliness. He clearly did not photograph well because I had always imagined him to be a tall, rather cadaver-like individual without a sense of humour. This short, dapper and vigorous little man soon had the hall in fits of laughter and was clearly widely respected, if not revered. Encouraged by his attitude, I generously forgave the fact that he had overlooked me in his general bonhomie and optimistically moved towards him to ask for a presidential favour. I was instantly bracketed by two very large, armed and mean-looking men, whom I might describe as more mature, but even less friendly versions of the Malawian Young Pioneers. The idea of presidential patronage died stillborn.

So, I returned to Malawi disappointed but not despairing. Once President Nyerere had left the hall, one of the least miserable of the officials walked back with me to my vehicle and pointed out that the refusal was all for the best. The southern coast of Tanzania that I was intending to research was, as he put it, 'filled with freedom fighters who would not make you welcome and the government does not have much control in that area.' *C'est la vie.*

I was not expecting to be back in Malawi on New Year's Eve, so at Chitipa I permitted myself to be persuaded by a Peace Corps volunteer, who I had met on the way up, to attend a school festivity. It changed my views on education and its benefits. The school principal got so drunk that he couldn't stand. Being a foreign scientist apparently carried a semblance of mystique, which resulted in every senior schoolgirl deciding that she had to dance with me. What was worse, in Malawi men dancing with men is apparently de rigueur, and I had problems fighting off two male nurses who wanted to share me with the girls. By the time I fled, claiming terminal exhaustion, it was clear that a break from all the driving and crowds was necessary and I decided that, as the Nyika National Park was barely 80 kilometres off the road,

I would make a small detour to visit this famous protected area. It was worth the effort.

The road to the park was good, and as one climbed to the heights of the escarpment, the scenery became more and more spectacular. Of particular interest to me was the rolling grassland which, at nearly 3,000 metres in height above sea level, looked almost identical to the rolling montane grasslands that surrounded me when I was game ranging at Giants Castle in the Drakensberg. The similarities did not end there, for Nyika was well stocked with game, and the first eland herd I encountered amazed me more than the grasslands.

The Cape eland living in the Drakensberg inhabit a mineral-poor region and they have certain characteristics that distinguish them from eland living in the lowveld of the Kruger National Park, Botswana and Zimbabwe. Calves are only very occasionally striped and females often have slender and, sometimes, very malformed horns. Lowveld calves are always heavily striped and the adults are much heavier and more robust.

The Nyika eland were virtually identical to the Cape eland of Giants Castle. The females showed the same characteristics of slender and malformed horns and although a higher percentage of the calves I saw had stripes, most of them did not. These external features were clearly a product of the highland environment, a discovery that I found particularly interesting.

The most surprising animal on view, just about everywhere, was the roan antelope. Until then, I had never associated roan with highland areas, so their presence was as exciting to me as the eland. There was an additional reason for this. When I was at Giants Castle, I had spent several years, in addition to my normal game ranger's duties, studying and recording San rock art in the dozens of rock shelters in the cave sandstone. During the course of my studies, I described the occasional paintings that were clearly of roan antelope. Roan paintings are, in comparison with eland paintings, extremely rare and many a vigorous argument has been entered into by both anthropologists and biologists over the possibility that roan once roamed the Drakensberg. Personally, I am a great believer in the view that many rock-art paintings are reflections of whatever took the fancy of the artist and that there is a historical component to our rich heritage of rock art. Many anthropologists feel that there is an overwhelmingly spiritual motivation behind the execution of the paintings and that they are not historical. I have no doubt that there was a considerable spiritual component, but that does not mean that there was not a simple artist's choice as well.

The roan in this Drakensberg-like environment added conviction to my supposition that, although there are no written records of roan having occurred in the Drakensberg, there is no ecological reason for them not having once lived there or having been at least occasional visitors. It is well known that the eastern escarpment in KwaZulu-Natal has been getting drier and cooler within the last 8,000 years or so, perhaps making the region less desirable to roan antelope, which prefer warmer habitats. These changes have taken place well within the known San occupation of the Drakensberg. It is, therefore, quite possible that the San and the roan once shared that beautiful area. The roan antelope on the Nyika, therefore, provided considerable stimulus to my imagination and gave me great joy.

On the second day of my visit, I drove over to the edge of the escarpment where a viewpoint situated close to the edge provides one of the world's great panoramas. The magnificent blue expanse of Lake Malawi spread out 1,000 metres below me, stretching north to Tanzania, east to Mozambique and south further than the eye could see. The green grassy slopes were artistically painted with outcrops of forest and everything was absolutely unspoiled. It probably has not changed since the Rift Valley was formed. I have never forgotten the wonder of it all. Wonder did not affect me alone, however.

The access road was very narrow and both coming to, and away from, the rather small viewpoint one had to drive with some care. After stopping, I left the Land Rover door open (placing the WWF logo, and 'Sea Turtle Survey' in full view of the road) when I walked over to the escarpment edge with my binoculars. I was barely conscious that another vehicle had come by and, unable to find space in the parking site, had driven on for a few hundred metres before coming to a stop.

I think that it was the vehicle stopping that brought it to my attention. To my surprise, it started to reverse back towards me, slowly and with care. Thinking that the occupant had decided that there was no reason why he or she should not share the site, I moved back towards the Land Rover, prepared to move on. The vehicle stopped right next to me in the middle of the road and the driver, clearly a South African, said, 'Forgive us our curiosity, but can you tell us what the hell a sea-turtle survey is doing on top of the Nyika Plateau?'

There was little else I could reply other than, 'Sir, you see before you a gross abuse of public funds!' There was a burst of laughter from inside the car and, without waiting for any further explanation, he said, 'Well, that's all right then' and drove away. I never saw them again.

While I am in confessional mood, I must admit that among my many faults is a passion for trout fishing and, strange as it may seem, the rivers on the Nyika, and the occasional dam, had been stocked with trout. I was aware of this and, as the trout had, like me, come originally from Scotland, I felt obliged to use them as an opportunity to cast a fly in some of the most beautiful surroundings in the world. Every evening I went out to the forested rivers, which were a serious challenge to the unskilled and skilled angler alike, and caught some small but memorable brown trout while being watched by reedbuck and zebra on the surrounding hills.

That was, I suppose, yet another abuse of public funds and, in the super-green environment of the twenty-first century, I expect there are those who would wish to have me burnt at the stake for daring to enjoy the presence of aliens in this unique African landscape. Personally, I remain quite broad-minded about such things, and it does not bother me now, and certainly didn't then, because in front of me was the long drive back to northern Mozambique to find out more about sea turtles.

CHAPTER 22

# The Morning Chorus and Other Excitement

The long drive back through Malawi and into northern Mozambique is almost best forgotten, but once I cleared the escarpment east of Lake Malawi I found myself in a countryside I had only read about in geography books. This was inselberg country, a gently undulating bush landscape punctuated by sharp granitic-rock intrusions that projected in excess of 100 metres out of the bush, each trailing a skirt of forest. It is spectacular and covers the whole far north of Mozambique right down to the coast.

The road as far as Nampula was tolerable and pleasant to drive on. After Nampula, however, as I turned north to Porto Amélia (now Pemba), the heavens opened and the road became a nightmare. There are lots of places where it rains heavily and this area may safely be added to the list. The dirt road was potholed so badly that the Land Rover fell into the holes with a splash and climbed labouring out of the other side. After 50 kilometres of this, I was exhausted. Land Rover in those days regarded windscreen wipers as optional extras and they never worked very well. The sealing on the doors was also suspect and the rain soon proved that it too was an optional extra. The inside of the windows fogged up and the windscreen wipers could not cope, so one crashed along, driving in the blind hope that one of the wet bits of the road didn't turn out to be a river.

At the first sign of civilisation, I pulled off and slept. As usual after such downpours, the next day was beautiful and I made good time to Porto Amélia. A few enquiries led me to the beaches and fishing villages and there, in considerable quantities, I found the remains of olive Ridleys. They were apparently very common in the region, along with green turtles and hawksbills. Even more importantly, they were reputed to nest, albeit in small numbers, along the coast. The fishermen informed me that the large green-turtle nesting areas lay further to the south on the offshore islands of the Primeiras and Segundas.

The local Makua people assured me that the coastal communities were all Muslim and that they were not allowed to kill adult turtles for religious reasons. Eating eggs was fine, but not adults. The number of carapaces that I found and measured in the villages from Porto Amélia southwards would suggest that the Qur'an had rather a weak grip on its followers in the region.

Arriving in a small village called Mossuril, a gem of a place along a coast that was a long series of gems, I was led to a German expatriate called Kurt Grosch. He had been living a hermit-like existence in Mossuril since before the Second World War and, as he made his living by collecting and selling seashells, he really was an authority on the coast and its biodiversity. Kurt is the sort of man that exploratory researchers like myself dream about. He found me a bed and I spent nearly a week in his fascinating company. I left richer in mind and spirit, and as the proud owner of a massive green-turtle carapace that Kurt had patiently dried. It was in beautiful condition and, having accompanied me for nearly 40 years, now hangs proudly in the staff bar at the Ezemvelo KZN Wildlife headquarters in Pietermaritzburg. Every time I pop in for a drink there, I am reminded of this remarkable man.

Kurt was despatched to Mozambique in 1936 by the Third Reich to establish, among the quite significant German farming community in the north, a series of willing supporters who could provide Germany with regular information and, in case of need, safe havens and succour. Kurt was no fan of National Socialism, however, and promptly dropped out of sight, becoming a beachcomber until the end of the war when he emerged from obscurity and quickly established himself as a malacologist of note. He maintained his lifestyle very adequately from the income derived from the shell trade, but, without any pressure to display success, he started and then abandoned many projects. His house was testimony to a person without pressure. There was even a yacht, partially built, in the garage, which, to my knowledge, never made it to the water before he passed away ten years later. This was probably because he realised that he would never be able to get it out of the garage without demolishing the building.

Kurt provided me with a broad-brush picture of sea turtles in the area. Their numbers had declined dramatically. He told me that before the war there were still fishermen catching turtles using remora fish, but the skill had since disappeared. Remora fish were caught, attached to a cord and released into the water when a turtle was seen. The fish would swim to the turtle and attach itself to the plastron by means of the unique suction pad on top of its head. Once it was attached, the fisherman would carefully and steadily draw the turtle close to the boat where it would either be grabbed by hand or harpooned.

Kurt arranged for me to go out to sea with local fishermen and provided me with invaluable contacts on Mozambique Island and among the local German sisal and coconut farmers, who, on many occasions, provided me with incredible hospitality. Ingrid and Henning Woermann in particular, with whom I stayed in contact for years, were not only incredibly helpful and hospitable to me, but also to friends who ventured into the north in those days. Kurt's advice was the same as that of the fishermen: go south to Antonio Enes and visit the outer islands where turtles nest in large numbers. So, I drove south to Antonio Enes, expecting to find Roger Oxley-Oxland, whom I had met in Lourenço Marques.

Roger, originally employed as a biologist with Anglo American, had visited the north to assess the prawn-fishing potential, fallen in love with the area, and a beautiful Portuguese girl called Julia, and had decided to enter the business himself. He raised the finance to buy his first trawler, *The Star of David*, and started to sell prawns. Since the *Star*, he had purchased more vessels and was ideally situated to help me to get out to the islands. It had all been arranged and I intended to spend a lot of time exploring the region. There was only one small problem – Roger wasn't in Antonio Enes.

However, some fellow prawn-boat owners were and I was quickly offered a trip out to the Segundas, the most northerly string of islands nearest to Antonio Enes. The *Belita* was due to fish in the region of Mafemede, the largest island in the string, which had once been home to the local caliph who controlled the region. The trip was extremely memorable.

We set sail about midday and several hours later dropped our first trawl near Mafemede. After an hour and a half the cod-end was brought on board, but with few prawns, so the skipper, a short, strong Portuguese, with a string of profanities handy for every occasion, decided to drop anchor outside the reefs of Mafemede.

We had prawns washed down with red Portuguese wine for supper and they were delicious. Along with superb Portuguese rolls, we had prawns and red wine for breakfast, lunch and dinner every day and, on later trips on Roger's vessels, the same haute cuisine persisted with the occasional change to squid boiled in their own ink. The latter was not my favourite as, strange to say, calamari had not yet appeared on South African menus. Prawns we knew because they were a real treat, but, I assure you that, after a week dining solely on prawns, they pale rapidly as a special dish.

I was offered a cabin and made myself comfortable as night fell, noting as I went below that every other crew member was carrying his mattress onto the deck. I did not join them, however, because we people of British stock know how to sleep with dignity and seek a degree of privacy during our resting hours. Once I was comfortably ensconced, I switched off the light and, I kid you not, was promptly joined by about 30,000 cockroaches (obviously of British extract), which swarmed all over me. I, mattress and all, soon joined the crew on deck and had quite a peaceful night's sleep. Until the dawn chorus erupted.

I used to be a smoker, and my father was as well, so the occasional cough in the morning to signal that the lungs still worked was not unknown to me. This collective bedlam, however, was the product of every other pair of lungs on board ridding itself of the previous day's accumulation of bad Portuguese tobacco, probably flavoured with cheap red wine and undoubtedly mixed with copious quantities of diesel. The *Belita* should have been in *The Guinness Book of Records*. It took over an hour before the noise of the sea around us could be heard above the coughing.

The captain very cheerfully shipped me ashore to Mafemede, where I interviewed some local fishermen. I was sad to hear that green turtles had once nested in large numbers on this group of islands, but were an extreme rarity now. I was assured, however, that there were still plenty of greens nesting on the Primeiras, the southernmost string of islands some 150 kilometres away. The fishermen had clearly been busy with other species, as there were racks of drying fish and piles of the most beautiful shells, obviously stripped from the reefs at low tide. It was painfully evident that no resource could maintain itself against such exploitation. It struck me that the shells and other resources would soon follow the green turtle of Mafemede into extinction.

Two days later, we had a fantastically successful trawl and the cod-end emerged from the sea to the applause of the entire crew. It was bursting with prawns; there must have been well over a ton of them. When they hit the deck, we were all busy for some hours, sorting the catch and packing the various valuable species into separate crates. So pleased was the captain that we headed back to Antonio Enes at speed, having collected more than enough prawns to show a decent profit for the trip. On arrival, the skipper, mistakenly believing that I came from colonial stock and expected to be treated with respect, had me carried ashore on the shoulders of a large and powerful crewman in lieu of the normal jump-into-the-shallows-and-walk-ashore technique. I did not even get my feet wet and it was done with such a sincere spirit of goodwill and friendship that I thoroughly enjoyed what was once a common practice.

The summer floods had swept away all the river bridges (they were rebuilt every year) south of Antonio Enes, so I went inland, once again travelling for some days. To my surprise, I found two friends from the Natal Mountain Club stuck in the mud in a Mini-Minor. George Zaloumis and Carl Fatti were doing their own adventurous trip around

Mozambique and looked fine, but their Mini was objecting to the abuse. I directed them to the Woermann's farm where, in due course, they were royally treated and their car repaired.

I regained the sea at Quelimane, but learnt very little other than that the Primeiras was the place to find nesting turtles. So, I drove inland once again, returning to Mozambique via Malawi (passing the beautiful Mount Mlanje en route) and via Tete to Gorongosa Game Reserve. Here, I met up with my ex-Natal Parks Board colleague Ken Tinley, now chief ecologist of Gorongosa, and his lovely wife Lyn, an artist of note, who graduated from the University of Natal in Pietermaritzburg the same year as I did.

Ken, like Anthony Hall-Martin, displayed a weakness for elephants and I joined him and his technician Costa in his eccentric Land Rover to observe a small herd of 16 animals feeding on the Pungue flood plain. Approaching them was painfully slow because the flood plain was drying out and we had to drive over hundreds of dried-out hippo and elephant footprints, which was extremely uncomfortable. When he deemed that we were near enough, Ken switched off the engine and we settled down with binoculars to record what the animals were eating. After about 15 minutes, I noticed with minor alarm that another herd had emerged from the bush behind us and was fast approaching. There were about 26 of them and they ignored us completely as they passed on either side to meet the resident group. A great deal of trunk waving and shrill grunts informed us that the second herd was welcome. We relaxed. We did not stay relaxed for long, however.

A large female suddenly lifted her trunk and trumpeted and the entire group of over 40 elephants all started trumpeting and writhing together in a compact mass. Ken said he did not like this (I agreed) and I noticed that Costa, a well-tanned Portuguese, appeared to be going green. It was at this point that the Land Rover's eccentricity became a real concern. Firstly, it had a habit of not starting, and, secondly, due to a damaged steering rod, its turning circle was three times larger than normal (for those who know Land Rovers, that is saying something). The herd was about 100 metres away when they charged. Ken uttered a prayer to a variety of gods as he reached for the key. I could say that

I considered it undignified to get out and run, but the truth is that I was riveted with fear to my seat. Ken tried to start the engine, but, as usual, it ignored him. He tried again and the engine started. The herd had covered about 40 metres by this time and was just getting into its stride. We were facing the herd and jolted forward, taking the hippo footprints as fast as possible and, because of the steering rod, we continued on a collision course for about 25 metres before we started to curl away from the enraged mob. It was one of the most frightening experiences of my life. When we regained one of the reserve roads, the nearest elephant was about 10 metres behind us and intent on doing us some serious damage. I have never been happier to exceed the park speed limit and we eventually left the herd behind. Costa had indeed turned green.

Ken, sensitive to the acute smell of fear surrounding him, decided that we had seen enough elephants for the day and took us down to the first hutted camp built in Gorongosa. It was abandoned, due to the water table that occasionally rose above the camp, and had been taken over by a large pride of lions whose members posed beautifully for us. It was a calming experience.

The next day, Ken announced that we would fly down to the coast over the Marromeu Game Reserve just south of the Zambezi mouth. We were going to undertake a buffalo count, but would detour along the beach so that I could check for nesting tracks of sea turtles. We took off in a Cessna 172 flown by a Portuguese pilot and I was soon lost in wonder as Ken explained about the structure and dynamics of the magnificent Pungue flood plain. En route to Marromeu, we flew over the most beautiful fan-palm (*Borassus*) country and then on to the flat grass-covered reserve itself. There were thousands of buffalo. I could hardly believe my eyes. We counted over 26,000 in half an hour. Individual herds often exceeded 2,000 animals and the grasslands were also dotted with modest-sized elephant groups and reedbuck. It was a sight that was, unfortunately, not to survive the various wars that were to follow in the next 20 years. A recent count suggested that there might now be 2,000 buffalo in the entire park.

Overflying the beach was a disappointment, however, as it was completely ploughed up by buffalo tracks for some 20 kilometres. It would appear that the buffalo spent a great deal of time on the beach, right up to where the first mangrove patches occurred near the mouth of the Zambezi. So, we returned to Gorongosa with no additional turtle information and the next day it was necessary for me to return to the beaches of Maputaland. After travelling through Southern Rhodesia, I reported my findings to Tony de Freitas in Lourenço Marques and arrived safely at Bhanga Nek, not entirely satisfied with my exploration of Mozambique.

A few weeks later, a letter came from Ken informing me that the Cessna, with the same pilot, had crashed (happily without any fatalities) the day following our flight. It seemed that Ken's appeal to the gods, when faced by charging elephants, had provided us with sufficient insurance for two days.

CHAPTER 23

# Of Rats and Mermaids

My second major Mozambique expedition in the Panda Wagon was somewhat better organised than the first, but it turned out to be nearly as unsuccessful. Late winter in 1970 saw me heading north on the most direct route to Antonio Enes. This time the roads were dry, the bridges were all either intact or rebuilt and it took me only five days to get to my destination. Roger Oxley-Oxland was in residence, as arranged, and plans were afoot to get me out to the Primeiras island group on the *Cobre*, one of his prawn trawlers. I had come fully prepared for a month-long sojourn on the islands but, of course, 'the best-laid schemes o' mice an' men . . .'.

The *Cobre* required repairs that would take five days after she returned to port, so Roger helped me to arrange some alternative research activity. I had undertaken to keep an eye out for dugong while searching for sea turtles. Although I had never seen one in the flesh, I had received reports of their presence around Inhaca Island in the south and the Paradise Islands near Inhassoro and that they were quite common near Mozambique Island and around Porto Amélia. Two years later, on my final expedition to the far north of Mozambique, there were also reports of dugong being very common. Known in that region as *mvua*, they were hunted widely and the meat was particularly sought after by pregnant women, who believed that it made childbirth easier. No use was made of the teeth or the hides. Fishermen, in all areas, felt that dugong numbers were declining.

In Antonio Enes there were numerous reports of dugong and also at least one professional dugong fisherman. Dugongs are large mammals, measuring up to 3 metres in length, and live off the rich beds of *Cymodocea* and *Halodule* that carpet the shallow warm coastal waters of the areas mentioned above. The magnificent estuarine system nearby, stretching 35 kilometres from the mouth of the Meluli River to Antonio Enes and over 20 kilometres wide at its broadest point, had extensive suitable habitat and I now had time to search for dugong.

Roger roped in a local pilot Sr Leitão and we had two extensive flights over the estuarine system in his small Cessna. Over two days of flying, we encountered six groups, totalling some 27 of these fascinating animals. This was one of the largest dugong counts ever recorded in Mozambique. While I was up in the air, Roger was tracking down the exact whereabouts of the professional fisherman and that evening we interviewed him and inspected his equipment. He used nets over 100 metres in length made of woven coconut fibre with a mesh size of some 60 centimetres. He said that he set them overnight in the narrower channels of the estuary. It is forbidden to kill dugong (or sea turtles for that matter) in Mozambique, a situation apparently unknown to the fisherman. He waxed enthusiastic about his craft, observing that there were fewer dugong than there used to be and that he was lucky to catch one or two a month. The available habitat in the river delta is enormous and, with so few animals, I was not surprised that he was only rarely successful.

As he was the only dedicated dugong fisherman in the area, it is likely that he was reluctant to provide the two of us with his trade secrets and I am sure that he could have told us much more about how and where he set his nets. From the air, we had seen feeding groups stirring up not insubstantial clouds of fine mud as they grazed the underwater grasses and, as the group progressed through the grasslands, long trails of suspended silt could be seen behind them. Such disturbances would have indicated to a hunter that there were dugong active in the area, and the size and spread of the silt cloud would have revealed the direction in which they were feeding.

## Of Rats and Mermaids

By the time I returned from the islands, Roger had persuaded the fisherman to lend us a dugong. The specimen was kept in Roger's deep freeze until I had inspected it and taken its dimensions. Dugongs are of the Sirenia family (so-called because early sailors [probably scurvy-ridden] believed that dugongs were mermaids or sirens) and are restricted to the warmer Indo-Pacific region. Inspecting the dead dugong, I could not help feeling that any sailor who mistook such a beast for a maiden had to be pretty far gone from one ailment or another, or literally mad with desire. It must have been rough going to sea in those days. This particular specimen was in very good condition without any visible damage, quite unlike the manatees (the dugong's American cousin) around Florida. Many of these carry symmetrical scars along their backs from passing outboard-motor propellers. The whole Sirenia family has, not unlike the sea turtles, suffered badly from overexploitation for the past few centuries, but there are definite signs of this situation improving.

Unfortunately, the conservation situation for dugong in Mozambique has not improved over the last 30 years, but the modest population appears to be holding on. At one stage, the global future of the dugong was seriously in doubt, but over the past few years there has been greatly enhanced awareness and protection of this shy and retiring animal, especially in Australian waters. Arabia and the Gulf states have paid a great deal of attention to protecting dugong, so the global survival situation is much better. With the declaration of more marine protected areas, Mozambique could quickly improve the status of its dugong populations.

At last, the *Cobre* was ready to depart and I loaded all my equipment and we set sail for the south. I immediately realised that I had blundered in buying 45-litre plastic water bottles. They are fine and light when empty, but heavy when full, and, what is more, they are flexible and awkward to carry, increasing the danger of an accident and their being burst. The little external taps, which I thought so useful, proved to be a real nuisance, as they were vulnerable to being snapped off and made the containers awkward to store. Just getting my water stores on-board involved an hour full of tension. However, eventually everything was loaded and lashed down and at midday we set sail for the Primeiras and Casuarina Island. My Robinson Crusoe adventure was about to begin.

It took two days and quite a few trawls before we dropped anchor off Casuarina Island during the early afternoon of 4 August. Out of the five islands in the chain (from the south these are Silva, Fogo, Coroa, Casuarina and Epidendron), I chose Casuarina because it was the largest, uninhabited (like all the others) and reputed to have the most nesting turtles. Casuarina lies some 15 kilometres off the mainland, measures about 1,000 by 300 metres at its widest point and is completely surrounded by white coral beaches. A magnificent coral reef stretches for at least a kilometre off any of the beaches. The terrestrial part of the island is dominated by casuarina trees and there is no fresh water except during rains.

Once ashore, I found a splendid campsite immediately behind the dunes in a sheltered site where all my equipment was quickly dumped by the crew of the *Cobre*, who were eager to be on their way. It was agreed that they would return to pick me up in 22 days' time. It was, therefore, with a small feeling of anxiety that I stood on the beach and watched the boat rapidly diminish and disappear. The coast of Mozambique was just visible on the western horizon and I was on my own without the ability to contact the outside world. I had no two-way radio and, in those days, no one had even heard of a cellular telephone. It was a little eerie.

The tent and flysheet were soon up, however, and the camp arranged to my satisfaction. By the early afternoon, I was off on my first patrol around the island to see what the nesting activity was like. My informants had told me that August was the peak nesting season in these islands, so it was disappointing to find that there were very few fresh signs of nesting. At that stage I did not know a great deal (actually, I knew virtually nothing) about green turtles and, although the peak nesting season of loggerheads and leatherbacks in Maputaland is November, December and January, there was no reason for me to doubt the information that I had been given. As it turned out, in later years, when I was more familiar with green-turtle nesting biology,

I would still have been confident that the information was reliable because green turtles at the northern end of the Mozambique Channel tend to favour the winter months for nesting. This is especially the case with the very active nesting grounds at Itsamia beach on the south coast of the island of Mohéli in the Comoros.

It may have been disappointing, but the beaches were beautiful, sparkling white and clean, so I was in no way worried about the lack of recent nesting pits. There were enough old pits to encourage me. In any case, it did not take long for me to encounter a more immediate problem, which took my mind off the turtles for a while. As I slowly made my way back to the campsite, I noticed a movement on the beach ahead of me. It was a rat wandering slowly down to the sea edge and obviously scavenging for food among the jetsam on the beach. A little further on, I saw another and as I approached they scampered, rather slowly I thought, back into the bush at the top of the beach. Roger told me that there were quite a lot of exotic animals on many of the islands – rabbits, goats and rats being the most common. Over the years, on my island visits, I came to encounter all of them, but this first encounter turned out to be the worst.

After it was dark, and I had prepared and enjoyed my first simple meal, I set off on my first night walk around Casuarina. During my entire visit, I would walk around the island three times per night. After the first week, I started to consider variation and would reverse direction from time to time because, although I found and tagged some nesting greens (a total of three), the numbers were a great disappointment. The one female, A508, became quite dear to me because she had real trouble nesting and came back on about five consecutive nights even after I had turned her, tagged her and tried to weigh her on the beach. All this sounds easy, but she did not go through all these insults without a fight. Once she was on her back, I leaned carelessly close to have a look at a barnacle on her plastron and she caught me a tremendous blow on the face that knocked me over the lip of the beach. It was some minutes before I recovered enough to come back and forgive her.

The attempt to weigh her came to nought. I rigged up a tripod of casuarina poles cut from the forest and attached my scale and rope to the top. Laying out a piece of robust fishing net, I dragged the old girl over it and then manoeuvred the net under the tripod. This took considerable effort as her shell was 111 centimetres long, which makes for one large green turtle. The general idea was that I would simply dig under the turtle, eventually removing sufficient sand to have her hanging free, and then it would be a simple case of reading her weight off the scale. By the time I had dug several tons of sand, and created a hole about 1.5 metres deep (and was exhausted), with no sign whatsoever of the stretch of the netting reaching its maximum, I knew I was beaten. For the record, I estimate that A508 must have weighed in the region of 200 kilograms.

The lack of turtles was unfortunately not matched by a lack of rats. When I came back to my camp the first night, it was swarming with them. Not one or two, but literally hundreds. The tent was full of rats, my stores were covered with them and the bush was quivering with their activities. I confess to receiving a real shock. One of the problems with being a dedicated reader is that one is exposed to experiences hardly likely to be encountered in real life. My immediate thoughts turned in panic to the fact that there were hundreds of rats, I was alone, I had no weapon except a speargun and they were clearly hungry. Ergo, the fate of the hero in George Orwell's *1984* sprang to mind – it was not a pretty thought. I had a vision of the *Cobre* returning three weeks down the line and finding a bare skeleton lying on my camp bed, slightly nibbled around the edges. What would they tell my mother?

With plenty of time on my hands, I rendered the camp almost rat-proof as far as food and equipment was concerned. I had several tin trunks and they withstood the depredations of the rats rather well. The same cannot be said of my typewriter, shoes, shirts, briefcase and anything else that was not hung from ropes in the trees. My briefcase carried the scars until I traded it for another some ten years later. I, fortunately, healed more rapidly and the few times that I was bitten did little more than intensify the tension. The main problem was the constant vigilance; everything had to be secured. I lost the most beautiful *Conus* specimen I had ever seen, complete with the container in which it was preserved in alcohol, and it took me three days to find one of my slops, somewhat the worse for wear, after it had been kidnapped.

The other problem was sleep. I didn't have much of it for the first week. The rats treated the tent as their own and, apart from scampering over me with gay abandon, quickly discovered the joys of skiing. They would scrabble up the roof of the tent underneath the flysheet, reach the peak and then slide with audible squeaks of delight all the way down the other side on to the sand. Once they got the hang of it, nights became quite a jamboree and I suffered grievously from a lack of sleep for the first few days. Then the war began.

My duty was clear: I had to rid the island of rats. A daily campaign of pursuing rats with my Zulu knobkerrie, which I had fortuitously brought with me, was initiated. As mentioned, there was no fresh water on the island, so the rats were in a constant state of water stress and thus had less stamina than me, even in my sleep-deprived state. Once caught in the open, preferably on the beach, a rapid pursuit normally brought the victim to bay within 50 metres, when I would administer the *coup de grace*. If among the trees, this required more skill, as the rats would dash around a tree trunk and then climb the tree from the other side. It was thus necessary for me to develop a blind round-the-tree-and-down blow never imagined by the maker of my knobkerrie. Even the baiting of rats with my rapidly diminishing stocks of cheddar cheese was undertaken and those brave enough to come within range were shot using my speargun. Various other less sporting techniques were employed and over time the rat population was reduced to tolerable levels.

Rats apart, Casuarina was a wonderful island. Not so much the land, despite the presence of a healthy population of Bouton's snake-eyed skinks, a lizard I had encountered in Maputaland on Black Rock, the only site in South Africa where this skink exists, but the reefs. They were completely unspoiled and I have never dived over a more beautiful site. The shallow *Cymodocea* beds had living harp shells, spider conches and red helmet shells in unheard of quantities. There were huge anemones, each with its own family of clown fish. The soft corals were covered in egg cowries, with a host of other rarely seen sea creatures. It was a paradise and, once in the 30-metre visibility blue waters off the outer reef, it was like swimming in an aquarium. Not being the bravest of souls, I was getting quite ambitious, swimming out several hundred metres, until I was sized up by a huge and really mean-looking sea pike. This incredible fish hung about me all the way back to the beach, disappearing the instant I pointed my spear at it, but reappearing again the moment I turned away. I had no idea how fast these fish could swim and I am sure that had he attacked I would have been in no position to escape.

Every paradise has its downside, of course. About a week before leaving, when I was talking to the trees (which is not really a problem unless the trees start to talk back) and in a generally relaxed mood, I saw in the moonlight in front of me what appeared to be a football. I leapt gaily towards it with the intention of giving it a playful kick, but, being barefoot, stopped by sheer luck at the final moment and switched my torch on. I was about to score a goal with what turned out to be the largest stone fish ever recorded. Prof. JLB Smith, in his wonderful work *The Sea Fishes of Southern Africa*, described the largest stone fish on record as being about 30 centimetres in length; this monster, with an abnormally distended stomach giving it the football shape, measured over 40 centimetres. Each of its 14 vertebral spines had a sac on either side with a centilitre of poison. Had I kicked the stone fish, I would certainly also have kicked the bucket. And that, I might add, is where it resides at ORI to this day. It is stored in formalin in the bucket in which I had kept my dried milk.

A more alarming surprise on the island was suddenly discovering that I was not alone. Ten days into my stay, I walked around the island to encounter, on the other side, a fishermen's camp with four fishermen who had called in for the night. Being a hospitable soul, I strolled in and greeted them like the country squire, forgetting for the moment that I was near a war zone, alone, without weapons (spearguns are used only in desperate situations) and unable to swim 15 kilometres at speed. Of course, as when competing in debates, all these brilliant insights only entered my mind after I left the fishermen and walked back to my camp. That night was a bit of a strain, but the following morning they arrived at my camp full of curiosity and listened to my radio for a while. They left with big smiles and I never saw them again. I try to believe the best of people, but in this situation I found myself appealing to the gods to ensure that this general practice was justified.

Dead on time, after 22 days, the *Cobre* dropped anchor offshore and I, along with all my specimens, goods and chattels, was transferred on board and we sailed north for two days. Unfortunately, it was dead into the teeth of a howling gale, which caused me to be seasick for the first time. Normally, I thoroughly enjoy rough weather at sea, but, perhaps because the trawler was being pushed with excessive speed for home and plunging dreadfully, my stomach gave up the unequal struggle to keep my prawns down.

Passing through the Gorongosa National Park on the way home, I was concerned to find that Ken and Lyn Tinley were getting worried about the approaching insurrection as Frelimo had established a forward base on nearby Mount Gorongosa and the war was spreading south. Once back in Durban, I, on the other hand, was heading east, for Madagascar.

CHAPTER 24

# Mozambique: The Far North

In 1972, with the inevitable help of Tony de Freitas, I was determined to complete my survey of the coast of Mozambique. The results of the first two expeditions had been published and I had undertaken to complete the survey as part of the grant agreement I had with the Gulbenkian Foundation. The final expedition would cover the zone between Porto Amélia and the Tanzanian border, then controlled by the Portuguese military who were deeply involved in the struggle with Frelimo. I anticipated some difficulties in getting permission to travel in the area. Once in Lourenço Marques, I was escorted by Tony to various departmental heads and military representatives, all of whom promised to arrange permission, but none of whom did so before I had to depart. Leaving Tony to make the final arrangements and obtain the written papers, I, once again, drove northwards on a wing and a prayer. In those days, getting anything in writing from the Portuguese at any level was like extracting teeth, unpleasant for all concerned and very frustrating. Tony had the patience of Job, however, and assured me that he would obtain the necessary clearances. How much do I owe that man?

The war had escalated dramatically since my last visit. Ken and Lyn Tinley had left Gorongosa National Park and were now resident in Lourenço Marques. The reported arrival of Frelimo fighters on Gorongosa Mountain had turned out to be true and there had recently been an attack on the main camp. According to Ken, the room I had slept in now had a fetching row of AK-47 bullet holes across one wall. This was, shall we say, less than encouraging.

Adding to my concerns was the presence, along the road north from Nampula, of increasingly frequent wrecks of vehicles (a catholic range, I might add) obviously blown up by Frelimo mines. In a Land Rover it is difficult to drive on tiptoe, but I tried my best and have never scrutinised a road surface more carefully. Halfway to Porto Amélia, I entered an extensive forested area, the shadows casting a real deepening gloom over the road. This heightened the tension, which suddenly went into overdrive as I observed a small horde of people rushing into the road and waving with excessive vigour. It did not take much imagination to believe that a wandering group of Frelimo were about to accost me and that this would be the end of my researches. Having no room to manoeuvre, I ground to a halt and was immediately surrounded by the excited mob, one of whom was brandishing a gun. I feared the worst.

One man politely greeted me and courteously requested that I do him a favour. I willingly agreed, in about five languages, with what I hoped appeared to him as restrained enthusiasm. He was delighted and cried out to a group hovering in the shade of the trees. They rose up carrying a litter on which was lying a local tribesman in a state of distress. He was covered in blood (still running, I hasten to add) and was rapidly deposited on the seat next to me, accompanied by cries of 'get him to the hospital at the next village'. I asked what had happened, in case the doctor in the next village needed to know. I must admit that I had a concern that he might not survive the trip and I was not sure whether my Portuguese was adequate to explain to the police what I was doing driving around with a dead body. It is amazing how clearly one's mind works in moments of panic.

'*Leao, leao, leao!*' ('Lion, lion, lion!') came the cry came from several quarters. This poor chap had been mauled within a few hundred metres of the road just moments before I came into view. Judging from the happy and concerned faces, I could not have chosen my moment of arrival more fortuitously. So, while the atmosphere was positive, I drove off with the poor villager bleeding all over my upholstery.

Genuinely concerned by now, I asked him if I could do anything, as I had a medical kit with me (which, I confess, seemed a little inadequate). He asked me for a cigarette, obviously having his priorities right, and I lit one up for him. He became quite loquacious and, although clearly in pain, seemed to accept the fact that lions are common in the area and that the attack was not personal. I admired his approach and hurried to get him to Alua, the next village.

On arrival, I asked directions and was eventually guided to a small square building about the same size as a modern-day freight container. There was no one there. Cries went out in all directions from my guides (within 100 metres or so, I had picked up a sizeable group of sympathisers) and, after about ten minutes, a uniformed figure came running up the road. This was the paramedic and his entire collection of paraphernalia was contained in a tiny compressed-cardboard suitcase similar to those used by pre-primary school children. He had me lay the victim down on the floor and announced, after I explained the circumstances, that I should leave everything to him. When he opened his suitcase, I could not help feeling that the lion was going to win in the end. The suitcase contained what appeared to be aspirin and a few clean bandages.

Pressing a packet of cigarettes into the hand of my ex-passenger, I left for Porto Amélia. My nerves were not taking this journey well, however, and they were justified in such a view. When I met up with the governor and his aides, to get my written permission to research the area, I was handed a letter from Tony de Freitas, which said, 'included please find your letter giving permission and calling for support.' Comfortingly relieved, I was horrified on searching the envelope to discover that Tony had forgotten to include the letter! It was a long day. The administrator then promptly put paid to my travels further north by vehicle. 'Too dangerous,' he opined, and promptly offered to arrange flights into the war zone for me, if my visit was approved.

It took two days before a shortened version of the permission was telegraphed from Lourenço Marques, but thereafter all was well. I was introduced to João Quintal, a cheerful, friendly and willing pilot and part-owner of the company Cadelco, who had a military contract to transport supplies to the outposts of civilisation still under the control of the Portuguese. The load, of which I was a part, consisted otherwise entirely of boxes of beer. It was one of the happier flights that I have contemplated taking. The plane was a Britten-Norman Islander, an amazing aircraft, able to land on very short and often badly maintained airstrips. I was asked to ignore the bullet holes in the one wing as João cheerfully assured me that occasionally a group of Frelimo would shoot at planes as they came in to land. Yesterday's incident was really an accident as they normally miss! I imagined us coming down amid a cascade of bursting beer bottles spreading joy to the entire countryside.

The coastline was fantastically beautiful. There were islands with lapping white beaches and occasional thickets of mangroves everywhere, dotting a sea of turquoise. This was green-turtle country par excellence. On the landward side, the country was less attractive and in the midst of the bush we flew over occasional fortified villages that were clearly heavily fenced and defended. It had been the policy of the Portuguese army to consolidate villages into organised and protected enclaves.

We landed without incident in Palma, a small and, to my amazement, totally unfortified village. The security fence consisted of one strand of wire, not even barbed, hung along the bushes surrounding the village. A welcoming party of a Portuguese captain and six armed soldiers awaited us. When I asked timidly about the security arrangements, the captain assured me that Palma was an open town and safe, as Frelimo came in most nights to buy stores. There was a sort of unwritten agreement that Palma was neutral territory. According to the captain, Palma was only 30 kilometres from the Rovuma River on the Tanzanian border and often received visitors from there. This was encouraging, until I was given my room that evening. Two soldiers, armed with automatic rifles, guarded my door all night. Perhaps the captain wasn't all that confident about his own propaganda.

The next morning, a clearly stimulated captain introduced me to my interpreter. He was a large, intimidating, but friendly Magonde with filed teeth and heavy tattoos in lines radiating from his eyes (sad to say, the week after I left he was detained as a Frelimo spy). The captain took the wheel of his vehicle, into which were crammed the six guards, the interpreter and me. He was genuinely interested in my work and fired question after question as we sped down to the coast along narrow sandy tracks. The captain drove rather too fast for the twists and turns of the track and, at an inauspicious moment, the vehicle lurched wildly out of a corner and he lost control. I watched with horror as we careered through a local cemetery, flattened half a dozen tombstones and then trisected a brand new grave, leaving three neat piles of newly created red mounds, each separated by a shallow trench with tyre-tread marks. Emerging from the cemetery, we entered a coconut plantation, narrowly avoiding a number of palms, before the vehicle stopped. The interpreter had gone white and so had I. In my view, demolishing graveyards did not rank high as a public-relations achievement.

Later, when we entered a coastal village, the Makua fishermen could not have been more helpful. They answered all my questions with conviction. Turtles and dugong were very important to the fishermen. They knew every species of turtle except the leatherback. Once again they claimed that adult turtles were rarely killed, because the people are Muslims, but eggs were eaten whenever nests were found. In one village there was a gutted green turtle on the beach, proving that Islam had a 'strong grip' on its adherents here as well. Killing had declined over recent years because of the war; they were frightened to go near the wilder parts of the coast because it was occupied by Frelimo camps.

Guerrilla wars are not a recommended method of conservation, but when in need one is grateful for whatever one can get. This was particularly so inland where large mammals such as elephant, buffalo and lion were thriving in the Niassa province south of the Tanzanian border. Neither side would fire at large mammals, however, in order to avoid giving their positions away. I am happy to report that the area is so rich in large mammals that the massive 42,000-square kilometre Niassa National Park, originally established in 1954, is now gaining a splendid reputation. The current manager, Clinton Rochat, is an ex-Natal Parks Board officer and a man really skilled in game capture.

When we returned to Palma that evening, I was in need of a beer and was amazed to find that with your ice-cold 2M beer came a free cooked *Scylla serrata* (mangrove crab). This is the largest coastal crab in the western Indian Ocean and, in my modest opinion, the finest seafood available. With it you were given a block of wood and a mallet to break open the huge pincers and get at the sweet meat within. What a treat, made more memorable because the sauce that accompanied it nearly took the skin off the roof of my mouth.

Delighted with the information gathered, I flew down to Moçímboa da Praia where the fishermen were equally helpful and then out to Mechanga Island, where I spent two days visiting local beaches and snorkelling on some incredible reefs. Mechanga was surrounded by a huge coral reef and, as the tide went out (sea level drops five metres along this part of the coast), jets of water emerged from the surface of the coral as the entire reef sank and compressed with the falling tide. In the sunset, these unusual sparkling fountains were a joy to observe.

Landing on Ibo Island in the Quirimbas group was frightening, as there appeared to be a large donga cutting through the middle of the airstrip, but the Islander took it in its stride. Whisked off by the local administrator, I visited the local villages and received information on a broad range of turtle matters, which was, alas, not all believable, such as 'nobody kills turtles'. On the way back to the strip, I was told that there was a prison for Frelimo soldiers on Ibo. He casually remarked that the prisoners were mainly Magonde, a tribe shared with Tanzania and famous for their woodcarvings, or *pao preto*, some of which could be bought at the prison shop. We called in and I was convinced that the Magonde are a unique class of woodcarvers. The artists obviously carve as individuals, each expressing what he sees in a piece of wood, and the work is beautiful. I purchased several items, one of which was an almost-finished carving of a man in repose, not unlike Rodin's *The Thinker*. This is an exquisite piece of work that I enjoy to this day. I left the far north with much relief as I had one more task to complete.

After I had travelled out to Casuarina Island in 1970, I was told that Fogo Island was actually the best nesting site for green turtles, but I had been unable to get there. With Roger's help in Antonio Enes, we arranged for me to go out from Moma, several hours' drive south, in a prawn trawler belonging to another company. After saying some very fond farewells to Roger and Julia, who could not have been more hospitable and never would accept a contribution towards meals or petrol, I set off for Moma. The next day, as arranged, I embarked on the *Vila Mona*, newly refitted with a brand new Volvo-Penta engine and off on its maiden testing voyage. What a stroke of luck. En route to Fogo we passed Casuarina Island and I almost wished that we could have landed. Fogo proved, from all the old nesting pits, that it was indeed a green-turtle nesting island, although the only recent nest we found was that of a hawksbill and two crewmen promptly robbed it of its eggs. My protests fell on deaf ears.

On the way back to Moma that evening, our arrival coincided with no fewer than nine other trawlers steaming into port. It was the most beautiful sunset and the whole flotilla was quite a magnificent sight as it swept into the estuary. The *Cobre*, my old transport to and from Casuarina, was in the group and, once anchored, the crew gave me a really warm welcome. We had a drink together that evening, bringing to an end my Mozambique research.

In 1975, as is well known, the Portuguese withdrew and all thought of conservation took a back seat in the country for 25 years. Most of the mammals in the large parks, like Gorongosa, were almost completely wiped out and the coastal and inshore waters were also overexploited. Many resources were severely reduced. Some 20 years ago, tourism began to pick up in Mozambique. Through Dr Scotty Kyle, a conservation colleague from the Kosi Bay Nature Reserve, we made contact with Pierre Lombard, a private tour operator who offered to start recording and tagging nesting sea turtles just north of Ponta do Ouro in Mozambique. Pierre has done excellent work every year since then and, at a workshop organised by the Mozambique Marine Turtle Working Group (with the financial support of IOSEA [the Indian Ocean–South East Asian Marine Turtle Memorandum of Understanding]) in Maputo in November 2010, Pierre was awarded the José João Award for Marine Turtle Conservation in Mozambique. Pierre has been quite inspirational and he is an integral part of the Maputaland turtle-study programme.

## Mozambique: The Far North

What is more, there is now a large group of enthusiastic formal and volunteer conservationists covering virtually the entire 3,000-kilometre coastline of Mozambique, devoting a great deal of time and effort to turtle conservation. The leader of this group is Marcos Pereira who, with his colleague Eduardo Videira, is doing an outstanding job of improving the future of sea turtles in that magnificent country. The Mozambique government has been making its contribution and, in 2009, it declared the Ponta do Ouro Partial Marine Reserve, putting some 80 kilometres of beaches, from Inhaca Island south to the South African border, under formal protection. Shortly before this, in 2002, the government declared the Quirimbas National Park in the far north, an area rich in sea turtles and dugong and so rich in memories for me.

One dream remains unfulfilled. The Primeiras and Segundas island chains remain unprotected. It has been announced that the Mozambique government intends to declare the entire string of islands a marine national park. When that happens I shall be a very happy man, but, in the meantime, it remains a work in progress.

CHAPTER 25

# Madagascar: Of Turtles and Tombs

On 4 August 1991, South Africa was shocked to learn that the *Oceanos*, a local cruise ship, was sinking off the coast of the Eastern Cape. The following day brought all manner of drama, not the least of which was the skill and bravery of the South African Defence Force helicopter teams who airlifted all the passengers to safety (the captain had already saved himself). What surprised me was that the ship had not sunk before, long before, when it was called the *Jean Laborde*, a French-registered vessel sailing regularly from Marseille to Madagascar. It was the ship that, in September 1970, carried me from Durban to Tamatave for the first time and gave me a nervous start to my researches in Madagascar.

To say that I was filled with trepidation would be an understatement. I had taken a course in special French at university and, having passed, felt myself adequately prepared. After 15 minutes on the *Jean Laborde*, however, I was terrified. I could not understand a word the crew were saying and when I spoke to them in what I thought was French they looked equally bewildered. I was about to be dropped off in a strange country, apparently without any means of communication, and had to survive there for four months. Little did I know what lay ahead.

The food on the ship was appalling, which caused a minor protest among the English-speaking passengers. A Mr and Mrs Smale became our leaders. Mr Smale was an economist, sent by the World Bank to assist Mauritius with its financial planning and was a delightful character. They said that their son, Malcolm, was keen on biology and he turned out to be just that. Dr Smale is currently the director of research at Bayworld in Port Elizabeth.

It was our collective opinion that the ship was not well cared for and, despite meeting numerous French expatriates returning from holidays in France and thoroughly enjoying their company, I was glad to disembark in Tamatave. I watched my Land Rover being lowered onto the quay and expected to be able to load my luggage and drive away. Alas, French bureaucracy is in a class of its own. It took two full days and a series of official forms, each with 16 copies, to clear my vehicle. All this despite having official papers from the government, both authorising my research and my presence, as well as an official invitation to attend an IUCN conservation conference in Tananarive, the capital.

Shaken, I eventually left the harbour and set off up the eastern escarpment encountering for the first time the constant stream of *taxi-brousses* that served the entire island as informal transport. South Africa, at that stage, had not yet developed our local equivalent, the minibus, so I was unprepared for the cavalier and cheerful drivers that served the people of this vast island. In truth, the only difference lay in the model of their vehicles. French cars, especially the Citroen *deux cheveaux*, were the taxis of choice in Madagascar and to this day I am amazed at the performance of these frail-looking vehicles. What is even more amazing is that when I was in Tananarive in 2011 for a turtle workshop, the same models were still hard at work.

The eastern escarpment is magnificent and the great swathes of traveller's palms that decorate the forest patches and hillsides are simply beautiful. It is small wonder that the Malgache have adopted the traveller's palm as their national emblem. This was only the first of a unique assemblage of plants and animals that were forever stimulating my interest as I travelled the length and breadth of the island.

The IUCN conference was noteworthy because of the conservationists I met and the contacts I made with local scientists. It was delightful to meet Sir Peter Scott again. I had met this famous bird artist and naturalist (and son of the famous explorer who died in the Antarctic) in 1961 at Giants Castle in the Drakensberg when I was a game ranger there. Peter had, however, been knighted since I last saw him on a memorable visit to the Wildfowl Trust in Slimbridge and I was a bit worried about how to greet him. He realised my confusion immediately and said, 'George. You always called me Peter; it is still the same.' He, in turn, introduced me to Charles Lindbergh, the first solo transatlantic pilot in the single-engined *Spirit of St Louis*, a warm and friendly man enjoying his twilight years as a conservationist. Later, after travelling down to Tulear (now Toliara) in the south, we spent a few wonderful days birdwatching and pursuing lesser flamingoes at Lake Ihotry near the west coast with a party including Drs Jean Dorst and Christian Jouanin (later to be so useful in my relations with Reunion).

My first real objective was to get to Europa Island, but the promised military flight from Tulear was delayed and I set off to research the west coast by myself. Turtles were common there and in earlier days the coastal waters must have been teeming with them. After travelling as far north as Morondava, some 400 kilometres over the most appalling roads and startlingly wide dry river beds, I was impressed beyond belief. The two dominant tribes along the south-west coast are the Vezo and the Sakalava. These coastal peoples are totally dependent on the sea and are incredibly well informed about sea turtles. Weather permitting, the fishermen go to sea every day in their outrigger canoes with their characteristic square sail, and they literally live off the sea.

Without wishing to digress too much from the turtle research, it is worth recording that the basic outrigger design was brought to Madagascar by the first Austronesian settlers when they crossed the Indian Ocean in the first millennium in their ocean-going outrigger canoes. They were the first humans to settle on this great island and they established a rice-paddy culture and a long tradition of fishing. They also brought their traditional way of ancestor worship, which I shall discuss later.

Sea turtles have always been, and remain today, an important source of protein for the coastal peoples of the west coast of Madagascar. Armed with my set of illustrative photographs and questionnaires, I visited several dozen villages. In every one I found numerous fishermen eager to talk about the turtles, or *fano*, and their value. The most common turtle caught, even today, is the green turtle, with loggerhead and hawksbill turtles being found more frequently than the olive Ridley turtle. Leatherbacks are almost never caught, although some fishermen did profess knowledge of them. Green turtles nested commonly on the mainland and on many of the nearby sandy islands, which lie within 20 kilometres of the shore. Hawksbills also nest on the Malgache mainland, but, happily for them, they are disparate nesters and do not congregate as many sea turtles do. This is undoubtedly the reason why there are still hawksbills in the waters of Madagascar. Strangely enough, from fishermen so well versed in sea-turtle knowledge, it was surprising to receive reports along many parts of the west coast that olive Ridley turtles spawn in the sea. This is, of course, not true, and in the far north of Madagascar there were reports of olive Ridleys nesting on the coast, although I did not personally see any such nests.

I estimated that at least 16,000–17,000 sea turtles were killed along these 400 kilometres of coast each year. Over the next few years, after completing my travels around some 80 per cent of the coastline of Madagascar, including some offshore islands such as Nosy Be, I concluded that about 40,000 turtles are killed in total each year. I would have thought that these numbers would have dropped today, but I have been informed by Blue Ventures, an active and quite admirable NGO, that their estimates of the turtles removed in the south-west are very similar to mine of 40 years ago. What has changed, however, is the attitude and customs of the local people.

During late 1970, when I first visited the south-west coast, every village had altars on the beaches facing the sea, which clearly illustrated the importance of the sea turtles and the respect that the fishermen had for them. Turtles were generally hunted at sea using nets and, commonly in the early days, with harpoons. There were two types of harpoons, the first being the *teza*, a single-barbed harpoon head of beaten iron some 10 centimetres long, which was loosely countersunk into a wooden shaft and, being attached to a cord, detached from the shaft when the turtle was struck. The *kisaisnitza*, used for smaller animals and fish, had three barbed iron prongs embedded in the shaft. This harpoon was hurled at the target and depended on the weight of the shaft to prevent the victim from swimming away. A similar harpoon to the *teza* was used in the St Brandon islands, but there the metal head was not barbed but simply nail-like. Once the harpoon head had been punched into the carapace of the sea turtle, the fisherman kept a strong tension on the attached cord, which was, apparently, sufficient to prevent the harpoon head from withdrawing.

No turtle was killed at sea, unless it was too large or too active and was threatening to swamp the canoe. Normally, all captured turtles were brought back to the village and carried up to one of the altars, which consisted of a bed of grass or casuarina leaves spread over the sand and surmounted by a wooden frame. Often, there were carapaces and plastrons on the frame and the heads of previous victims were impaled on a stake adjacent to the altar.

The turtle was never killed out of sight of the sea. Fishermen believed that this would have shown disrespect for the animal. The spirit of the turtle, having sight of the sea, could choose its moment to return to the water. The turtle, therefore, according to the belief of the fishermen, was never actually killed by them. As they dismembered the turtle, its spirit would select its own moment to depart the body. Personally, I considered the killing process distressing, as they cut the animal up alive, but when I said so most fishermen simply shrugged their shoulders, obviously feeling that a *vazaha*, a foreigner, couldn't be expected to understand these things.

Apart from the shell and the head, all the parts of the turtle were eaten or sold. The Vezo people ate the heart because it promoted long life. This is because sea-turtle hearts, like all reptile hearts, tend to beat long after the rest of the animal is dismembered, giving the appearance of a remarkable ability to survive. In northern Mozambique, by contrast, a turtle's heart is not eaten because it is believed that it prevents death.

Hawksbill turtles still provided a steady supply of tortoiseshell to the tourist and export markets in 1970 and tortoiseshell products continue to be sold as tourist curios to this day. In the case of the hawksbills, however, there is no doubt that the population has declined dramatically over the past century.

## Madagascar: Of Turtles and Tombs

During 2011, when discussing sea turtles with some Vezo tribal leaders in Tananarive, I was horrified to learn that, over the past 40 years, the traditional harpoons have disappeared. Even sadder was the fact that the system of altars and almost ritual slaughter of sea turtles had also disappeared. The staff members of Blue Ventures and other NGOs active in the region had never seen such things and were delighted to see the photographs I had taken in 1970. In truth, with the persistent growth of human populations and the gradual homogenisation of customs and appearance, the diversity that once was such a phenomenon of human society is slowly eroding. One can only reflect on what a privilege it was to have seen some of this island culture when it was in full flower.

Madagascar was unique in that if one wanted to see diversity, quite apart from the animal and plant wonders, the Malgache ways of dying displayed some of the greatest variations on the planet. There are 26 tribes in Madagascar and each tribe appeared to have its own way of treating its dead. However, just as the way of treating turtles has changed, so too are the customs of respecting the dead.

Fundamentally, the practice of ancestor worship came from the Far East with the first settlers and spread, with variations, throughout the island. When travelling along the coasts and through the various tribal regions, I could not help but wonder at the various ways the dead were disposed of. This could vary from a single stone standing upright with a *zebu* (Madagascar's unique species of cattle) skull and horns attached to the top, which represented the death of a tribal member far away from the home village, to massive structures 2 metres high, 10 metres long and 10 metres wide, surmounted by dozens of skulls and horns of *zebu*. The number of skulls apparently reflected the wealth and influence of the deceased. The Antandroy tribe erected 10-metre high slabs of stone, the more recent ones beautifully symmetrical and obviously worked by skilful stone masons, and incorporated them into the walls of their tombs. The Mahafaly people decorated their large tombs with carved wooden totem poles called *aloalas*, each about 130 centimetres tall and topped by an astonishing array of carvings. Each *aloala* depicted a highlight of the deceased's life, for example, that he could write, had been arrested, could shoot, owned many cattle and could ride a bicycle or drive a car. Few graves in the world offer the entire biography of the occupant as does a Mahafaly grave.

The more modern graves of the Vezo are all concrete and whitewash, decorated with a bizarre range of paintings depicting, in colour, what the Mahafaly portrayed in carvings. Of the coastal burial sites, however, the old Sakalava graves near Morondava on the west coast are to me the most memorable. A visit to this cemetery demands that one reviews, entirely, one's beliefs in what a cemetery is there for. What the Sakalava graves lack in volume (each is no larger than the common grave in a European cemetery), they more than make up for in decoration. Each grave has a wooden palisade, at each corner of which is a large wooden carving representing a sexual theme. This can be a simple carving of a glossy ibis, apparently representing fertility, to startlingly pornographic carvings depicting the size, skill and endurance of the deceased, either by him or herself or in glorious copulation with a member of the opposite sex. It was truly a sight to remember.

And that is just about all that remains of this practice. Even when I was there in the 1970s the practice was declining and the newer graves were simpler with less noteworthy decorations. All the carvings had originally been painted in brightly coloured oil-based paint, of which little remained, having been undermined and flaked off by termites. One saw a clear progression of graves as one walked through the cemetery, from the new, surrounded by brand new wooden palisades, to mere piles of wooden fragments gradually being dispersed by the termites. In between, however, were carvings ravaged by the termites that still maintained the pure form of their original subject. In that form they presented the real art of the sculptor and became most beautiful before succumbing to the termite onslaught. After departing Morondava, I wondered how long this particular art form would persist and how long it would take for the termites to destroy what must have been one of the world's most interesting burial practices.

In the highlands of Madagascar, the Imerina tribe builds more robust stone, brick and mortar tombs in which the dead are interred, being wrapped in a cloth called a *lamba*. Each family builds a tomb commensurate with its wealth and many deceased are kept therein only to be removed from time to time to be rewrapped, introduced to the extended family and, after a riotous celebration, replaced with due honour into the tomb. The French name for the ceremony is *retournement* (the returning) and the Malgache name for the tomb is *fasana*. An excellent treatise on these practices, *Placing the Dead*, was written by Dr Maurice Bloch.[1] It is well worth reading.

At a house party in Tananarive in 1970, Maurice scorned my attempts to get him to accompany me to a *retournement* and I, being unaware of the extent of his expertise, thought that was very narrow-minded of him, and said so. That crass observation earned me a real raspberry from our host, the British consul Mr Gerry Warder and his wife Betty, who tartly explained that Maurice had just spent two years doing his PhD research on the subject and was 'tombed out'. I blush to this day.

Madagascar is a declining asset in so many ways. On my three working visits, I drove nearly 8,000 kilometres in my Land Rover, surveying most of the coast and criss-crossing the island several times. The unique wildlife is fast disappearing on land, concomitant with the felling of the forests for slash-and-burn cultivation, as the ever-growing population of Malgache people strive to find the means to survive. The sea, to the tribes of the west coast, represents their sole source of economic benefit and the sea turtle is an important component of this. The fact that sea turtles have been an even more important component in the past is obvious in that fewer and fewer records exist of turtles nesting on the mainland. How, then, is it possible that today, in the twenty-first century, sea turtles, especially green turtles, are still making a significant contribution to the lives of the Vezo, Sakalava and other coastal tribes when the coastal nesting areas are nearing extinction?

The answer lies in the oceanic islands surrounding Madagascar and that was where my researches were taking me.

1 Bloch, Maurice. 1971. *Placing the Dead*. London: Seminar Press.

CHAPTER 26

# Europa:
# A Superfluity of Animals

Nothing had prepared me for Europa Island. Named after a British surveying vessel in 1874, Europa is situated in the Mozambique Channel more or less halfway between Madagascar and Mozambique. By local standards, it is quite a large island of raised coral, being 25 kilometres in circumference with 5,000 hectares of dry land and a huge lagoon. Encouraged by the occasional Victorian-era expedition report that stated *'tortues sont abondant'* ('turtles are abundant'), I eventually made my way there. It had not been easy to get this far because the French in those days were almost paranoid about keeping research on their territory in the hands of French scientists. Although I had originally been granted permission to visit the island and to travel there with a French chartered aircraft, I waited in vain for the plane to materialise. After several weeks' delay, during which time I researched up the west coast of Madagascar (see Chapter 25), it eventually arrived on 5 November 1970. Accompanied by a local French scientist from Tulear, M Andre Maugé, we took off in a twin-engined Dakota for Europa. In order to get back, however, I had to charter a twin-engined Aztec from Air Madagascar, which, after some two weeks' delay, eventually landed us back in Tulear on 19 December.

Europa had once been settled by some Seychellois and a small graveyard marks the passing of these courageous pioneers. It was never an easy life there, without a source of fresh water. Coconuts did not do well. The mass exploitation of nesting turtles eventually reduced the nesting population to economically unviable levels and the last desperate attempt at a cash crop, sisal, also failed. All turtle killing was stopped by legal decree in 1923 (although the piles of turtle bones from the period were still visible in 1953) and the island was abandoned. In 1954 the Meteorological Service of Reunion established an invaluable weather station there.

Armed with 400 tags, I was ready for anything, but not for the sight of hundreds of green turtles swimming in the clear blue waters off the west coast of the island as we circled it en route to landing on the original airstrip in the south. We were met by one of the local meteorological staff, the only residents on the island at that time. In later years, as the meteorological equipment was upgraded and automated, the staff became redundant and disappeared. They were replaced by military and police personnel who were ready to protect the island from invasion by Madagascar, which, after independence in 1974, claimed all of the scattered islands as its own.

We were made very welcome and I could not wait for the evening as the beaches beyond the meteorological station were pockmarked with turtle nest pits. I was eager to work with my first green turtles, so when darkness arrived I kitted up with my usual summer working attire: a pair of shorts and my bag with tools and tags. Having noticed quite a lot of metal debris on the beach, I decided to don a pair of slops. As I approached the door, my colleagues smiled broadly and asked, with some amusement, whether I was going outside dressed as I was. When I replied in the affirmative they cried, '*Bonne chance!*' and waved me on. It took me precisely ten seconds to realise why my companions were smiling: I walked outside into a wall of mosquitoes, not just thousands, but millions. This species, *Eretmapodites plioleuca*, is rather a unique mosquito that neither carries nor transmits any diseases, but is content merely to drink you dry. Totally unsophisticated in their enthusiasm, and obviously driven by the first-come-first-served approach to survival, they simply blanket every exposed space and sting viciously. My near-naked arrival was clearly a gift from the gods and to say that my rapid reappearance in the station was ignominious would be an understatement. It was, however, a question of losing face or dying.

Shortly thereafter, I went back out covered from head to foot with socks, shoes, gloves and a head net identical to that worn by beekeepers. I must have looked like a man from outer space and, at 32° Celsius, I was sweating before I left the door. The nesting turtles were, however, marvellous. There were hundreds of them crawling all over the beaches and I was faced with the brilliant prospect of possibly using, on my first night, every tag that I had brought. With an anticipated month to go, I rapidly reassessed my game plan (as the rugby fraternity would have put it) and, instead of the planned saturation tagging (even today it makes me smile to think about that), I decided to mark out the nearest beach into 100-metre long stretches and tag animals nesting only within the demarcated zones. As it turned out, I finished all of my tags within three weeks and, as a result of a passing cyclone, had another three weeks thereafter to complete my assessment of the nesting numbers without a tag in hand.

This is all very embarrassing to record now, but, as has been mentioned elsewhere, one is guided by one's own experiences. To me the word *abondant* meant that there were lots of turtles. To me, after five years on the Maputaland beaches, 'abundant' represented about 30 turtles encountered every night. The obscene mass of animals pouring out of the sea on Europa Island was staggering and I was thrilled. All over the world, mass turtle nesting of this magnitude had reportedly disappeared. But, here it was, one of the world's greatest biomasses, spread out before me and shared only by several million mosquitoes. What a night. I sat there, absolutely awestruck, with females coming from and returning to the sea on either side for several hours. This was the turtle equivalent of sitting in the middle of the wildebeest migration on the Serengeti. It was an experience never to be forgotten.

The mosquitoes made sure of that. Every time I bent or knelt to tag a turtle, some part of my anatomy became exposed and, within a split second, I would be squealing and writhing to try to cover the vulnerable spot. The night was rent by my odd cries and I drew on language that my mother had desperately tried to rid me of. One really has to experience this insectivorous onslaught to believe that, if one were exposed all night, one would be dead in the morning.

Once the first thrill of the evening had worn off, and I had started to get to work on the turtles, I realised that I was not alone in suffering from mosquito depredation. As the huge females emerged from the sea, they were immediately blanketed with mosquitoes. Every crack separating the scales on the head was clearly outlined with several centimetres of mosquitoes trying to get at the softer, more vulnerable skin between the scales. The same thing happened to the skin around the eyes and even the tortoiseshell carapace scales were outlined by piles of mosquitoes.

One gets used to anything, of course, and in the face of the turtle wonders the mosquitoes eventually became a secondary problem. Before leaving them, however, it is perhaps worth noting that, except during rains, there is no fresh water on Europa. The mosquitoes have, as a result, adapted to laying their eggs in slightly saline water. Around the lagoon are thousands of fiddler crabs, all of whom dig holes in the mangrove mudflats. These holes are normally flooded by lagoon water, which is much saltier than sea water. When it rains, however, fresh water runs into these holes and comes to rest on the surface of the hyper-saline water, forming a lens of less-salty water. It is in these fortuitous lenses of weakly saline water that the mosquitoes have adapted to laying their eggs. Thus, with the coming of the summer rains, comes the frantic breeding of the mosquitoes. This coincides with the peak nesting periods of sea turtles and sea birds, which, need I add, are accompanied by turtle researchers.

This being my first experience with green turtles, I did not realise that they would be different from those in other parts of the Indian Ocean or, indeed, in other parts of the world. When I had completed my studies on the other scattered islands, however, and was comparing the fine morphological details of each turtle population, I noted that the Europa greens were larger and rounder than the other populations. The largest female that we weighed, with some effort, was over 210 kilograms. Imagine my delight, in more recent years, when other biologists, notably Drs Damien Broderick and Delphine Muths, more skilled in genetics than I, described Europa green turtles as having their origins in the Atlantic Ocean. The Europa population is genetically unique in the Indian Ocean. Genetic differences between other green-turtle nesting populations in Tromelin Island, the Glorioso Islands, Mayotte, Juan de Nova and Mohéli are far fewer and they are all clearly of Indo-Pacific origin. Delphine and her colleagues, such as Dr Jerome Bourjea, have demonstrated that each nesting population has unique genetic characteristics. The remarkable feature of these closely allied nesting groups is that, although females from each population share overlapping feeding grounds all round Madagascar and along the east coast of Africa, there has been only one record of a female nesting on two separate islands. Many years ago, a green turtle, tagged on Tromelin Island on the eastern side of Madagascar, was found nesting on Europa and some years later was recorded nesting back on Tromelin. This is so unusual that even I, optimist extraordinaire, find it hard to believe that it might not have been a recording error or duplicate tag (such things happen even in the turtle world). Then again, occasional females get lost and are forced to lay their eggs on unknown shores and from such accidents new turtle nesting colonies are established.

From the 400 tags used came recoveries from Mozambique, South Africa and Madagascar, and tags put on by my successors on Europa demonstrated that this nesting population was distributed all over the south-western Indian Ocean. Europa is, without doubt, the most important green-turtle nesting colony in the western Indian Ocean. It is estimated that between 15,000 and 18,000 females nest there per year. Quite unlike our loggerhead and leatherback nesting populations in Maputaland, green turtles nest throughout the year on Europa. When I visited Europa in June of 2010, I was delighted to note at least 25 females nesting per night. What is even more exciting is that Stephane Ciccione of Kelonia has assured me that, as a result of their constant monitoring, it is evident that the numbers are increasing. The nearest nesting colony of equal, if not greater, importance is the south coast of Oman, the beaches of which host tens of thousands of green turtles annually.

In years to come we shall get even more accurate estimates as, on the 2010 expedition, Stephane and I marked out discrete sections of beach, using GPS instruments for accuracy, with aluminium signs for use by the local gendarmerie, which agreed to monitor the nesting crawls every day in future.

When I first visited Europa in 1970, I was disappointed to find that residents had taken little care of waste disposal over the years. There was rubbish lying everywhere, not only on the beaches, and within a short drive of the meteorological station parts of the island were simply rubbish dumps. I had noted this with some sadness when I submitted my first report to Reunion and the response was instantaneous. The chief of the Meteorological Services, M Marcel Malick, immediately ordered a total clean-up and strict rules were instituted for all temporary occupants of Europa. Today the island is spotless. Even the local gendarme rides a bicycle round the beaches every morning and picks up plastic bottles and other debris, which he brings back for regular removal by military aircraft to Reunion along with all other waste generated by the military. The French management of Europa, now a fully protected nature reserve, is exemplary.

The spectacular numbers of adult green turtles during the summers I was there showed that such large numbers have their disadvantages. For starters, there are many beaches accessible only at high tide. As the tide falls, nesting females, on laying their eggs, have to negotiate the raised coral platforms, or *champignon*, now obstructing their route back to the water. Many fall into holes and perish, being unable to extricate themselves, while others, even if surviving until the tide returns, die as the heat of the sun disorients them. Other turtles arrive in the early morning to lay, and by the time they have finished are overheated and, losing their way, wander inland from the beaches to die in the bush. A serious problem of having too dense a nesting population on a finite number of beaches is that hundreds of nests are destroyed each year by nesting females digging up previously laid eggs. This is a bonanza for scavengers such as turnstones, crows, hermit crabs and the introduced black rat.

The benefits for other animals of several million turtle hatchlings emerging every year should not be underestimated. Different predators have different techniques for making use of this food source. Any clutch of hatchlings emerging during the day, generally as a result of it being cooler, heavily overcast or raining, is completely wiped out within minutes by the ever-present frigate birds. Most, if not all, of the clutch will be gone within 25 metres of the nest. Those that make the water are simply scooped up with great skill as they rise to the surface to draw breath. Once the feeding frenzy is at its height, frigate birds will even scoop a hatchling out of your hand.

The hermit crabs, of which there are thousands on the beaches, each about the size of a tennis ball, have a collective technique. They operate at night, the more normal time for hatchlings to emerge and make their rush for the sea. Of all the species of turtle hatchlings that I have watched emerge and take their first steps to adulthood, the green turtle is the most active. It is capable of a rapid run, its little flippers flapping frantically, and with good reason. It has to run the gauntlet of hermit crabs. The crabs, in their turn, as soon as they spot a group of hatchlings, hitch up their shells and run at a hatchling. Catching up from behind, a crab grabs a hind-flipper and then lowers its own shell onto the beach, forming an anchor too heavy for the hatchling to drag. Within seconds, a horde of other crabs descend on the duo and pile up into a heaving bundle, all trying to get a share of the luckless hatchling. During the peak nesting period, the beach appears to have a large number of footballs scattered across it, each a rapacious heap of hermit crabs. Such concentrations of crabs leave clear lanes on the beach through which hundreds of fortunate hatchlings sprint safely to the sea, thanks to the sacrifice of their siblings.

Once in the sea, hatchlings have to negotiate a minefield of groupers, black-tipped reef sharks, as well as their larger cousins offshore, and a hundred species of marine predator. It is amazing that perhaps two of every 1,000 hatchlings making it to the sea survive to adulthood. Carried away by local currents, the green-turtle hatchlings will spend a year as pelagic predatory wanderers. They will then seek a more sheltered feeding ground with abundant food resources, in the form of marine angiosperms or algaes, where they will settle for some years. Europa lagoon is one such sanctuary and there are hundreds of plate-sized young green turtles feeding off the copious plant reserves on the bed of the estuary.

It is in this lagoon that, for many years now, my colleagues from Kelonia and Ifremer in Reunion have been catching young greens on the islands by using the rodeo method perfected by Col Limpus in Australia. Simply put, in the shallow waters of the lagoon, one pursues turtles using a rubber duck with a diver in the prow. His or her job, at the crucial moment, is to dive from the speeding boat and grab the turtle. It is not as easy as it sounds, however, and when my turn came in 2010, at 71 years of age, I was reluctant to try this unless the turtle was clearly fleeing over a sandy bed. The thought of landing with my head and hands in a coral patch held little attraction for me. At Juan de Nova the week before, Stephane, with admirable but foolish dedication, dived for a turtle and emerged looking like a terminal haemophiliac with blood pouring from several large indentations on top of his head. Need I say that he was clutching the turtle? Being a considerate friend, Stephane skilfully guided a chicken turtle into an appropriate sandy patch and over I went, earning myself a diploma from the Federation of Flying Turtle Catchers of Les Îles Éparses – veteran class! I was in good company as similar diplomas were earned that day by Anna and Nicholas Tisne. They were our hosts on the 28-metre yacht *Antsiva* on which we spent 24 days, leaving Nosy Be and working on Juan de Nova and Europa, before disembarking at Tulear.

A number of chicken turtles in Europa are now carrying satellite tags and data loggers and as each year's catching season passes, my French colleagues are gaining deeper insights into the early lives of green turtles.

Europa is host to many other species of plant and animal, and many endemics have been recorded. The birds are fantastic. There are hundreds of red-footed boobies, frigate birds (two species), up to a million sooty terns in the nesting season, straw-tailed tropic birds (a unique subspecies), the beautiful red-tailed tropic bird and many others. Amid this kaleidoscope of species there are the introduced aliens, such as the black rat and also a black goat. Goats were introduced by an early French explorer in 1860 and have survived with remarkable facility.

The population of goats is quite large and healthy-looking and, I am not ashamed to say, very tasty. In 1970, when there was not a regular military supply plane arriving, meat was rare. Once a week, one of the cooks and some volunteers (I was much more fleet of foot then) would sprint into the bush, running down youngish kids. May I say that this was nearly as dangerous as the turtle rodeo because Europa is basically an uplifted coral reef covered in xerophytic vegetation designed to withstand a lack of water. Every grass species and tree is drought resistant, hard and unsympathetic. Coupled with the fact that we were sprinting over heavily weathered coral outcrops, we never returned to the kitchen unscathed, but we never came home empty-handed either.

The secret of being able to catch a goat on foot was thanks to the general lack of water. The goats lacked stamina and appeared to be in constant nutritional stress. This was clearly illustrated one day when I foolishly decided to walk right around the island, checking some of the less accessible beaches for nests, and neglected to take any water. By the time I had to turn back, I was beginning to flag and two things happened to illustrate what a lack of water can do to man or beast.

Some 5 kilometres from the station, I was walking along a piece of track covered with sand and found myself gradually gaining on a goat walking in the same direction. The goat was clearly suffering from the heat as much as I and did not become aware of me until I placed my hand on its back and said hello. It fainted. After dragging the beast into the shade of a nearby tree and apologising profusely, I walked on. Some kilometres later, I was hallucinating, seeing in front of me a can of Heineken beer, sweating cool droplets, which made me think of Sir Pellinore questing for the Holy Grail. The strange thing was that I had stopped drinking beer in 1957 (having tried it perhaps over zealously when I left school) and had brought none with me to Europa. By the time I reached the station, I was staggering and lurched past my colleagues to the fridge where, wonder of wonders, there was indeed one can of Heineken left. Without thinking, I drank it, earned the rage of Andre Maugé (it was his last) and launched myself back into a lifetime of enjoying a cold beer on a hot day. My inconsiderate action was never regretted and beer has accompanied me to many other islands, only to prevent heat stress, of course.

Europa has been my Shangri-La. During those six weeks, I improved my French enormously, thanks to my patient colleagues and a massive supply of French comic books, including the inimitable *Asterix*. This amazing comic book taught me much about the French sense of humour; I have always given credit to the little hero for his help. When we dropped anchor off Europa in June 2010, my French colleagues presented me with the latest *Asterix*, in French of course, suitably inscribed 'In recognition of your first visit and the gifts of *Asterix* 40 years ago'.

CHAPTER 27

# Tromelin Island

'Everybody look for the island!' called the pilot of the Dakota as, two hours after leaving Reunion Island, we were still flying over a featureless expanse of Indian Ocean. Having absolutely no idea what to look for, I peered out of one window while Garth Batchelor, my colleague from ORI, who joined me on this leg of my research, peered out of another. Glancing around the plane, I noticed that every occupant was glued to a window and I am sure, apart from the pilot, they were all first-time visitors and probably as ignorant as me. When the pilot spotted Tromelin, we were all quite taken aback. It appeared to be all airstrip, with a beach at one end and a rocky shore at the other. It was only 1,700 metres long and 700 metres wide, quite flat, with a high point of 6 metres above sea level. The level surface was dominated by the meteorological station and the local staff cheerfully informed us that the entire island could be swept over if a cyclone struck it directly. This was encouraging as the cyclone season was near to starting and Garth and I were to spend nearly a month here tagging green turtles.

Tromelin is 350 kilometres east of Madagascar and 500 kilometres north of Reunion, a very remote spot indeed, but was reported to have a healthy nesting population of green turtles. We arrived on the island in October 1971 and, as it turned out, we were a little early to catch the main flood of nesting females, which arrive in late November, December and January. Nevertheless, we managed to tag 35 large green turtles and collect a mass of data on adults, eggs and hatchlings, which was more than sufficient to compare this population to the other major nesting beaches in the south-west Indian Ocean. Later recoveries of tagged adults showed that, after nesting, this remarkable group scattered across the region. The females returned to feeding grounds off Mauritius, Reunion, the St Brandon islands and to both the east and west coasts of Madagascar, intermingling with females tagged on Europa and other colonies. My estimations of the total nesting population of Tromelin, at between 300 and 400 females per year, turned out to be pessimistic. Over the next decade, G Batori, the first of the series of young Frenchmen who followed me, having spent much longer on the island, established that there were some 3,000 females nesting on this kilometre of beach, which was only accessible by swimming hundreds of kilometres across open ocean. Given the small size of the island, and the fact that it was the exposed tip of a volcanic seamount rising 1,000 metres from the ocean floor, the successful arrival of nesting females proved, once again, what astonishing navigational skills are possessed by sea turtles. This really was the equivalent of finding a needle in a haystack.

When Garth and I were on Tromelin, four meteorological staff from Reunion were working there. Today, with the work having been automated, the meteorological staff members have gone. We were welcome company and were often joined by them when tagging turtles, digging up eggs or catching hatchlings. Even more important to us all was trying to reduce the numbers of black rats on the island, as they were a constant threat to everything, from turtle hatchlings to birds. Tromelin had more than its fair share of aliens as the black rat, the common house mouse and the rabbit subsisted on the natural resources of the island.

The rabbits prevented the ground vegetation from reaching any size at all and, therefore, maintained an almost bare surface over the plain of the island, which undoubtedly had a drying effect on the environment. Situated, as it was, so far from anywhere, its vegetation must have struggled to get a foothold. Once the rabbits were introduced, it is very possible that some very rare, if not unique, species were rapidly lost. Following the negative comments in my report to Reunion, a massive effort to remove aliens was implemented and the rabbits have all gone and the rats were reportedly extinct, but I doubt that.

In 1971, however, the rats were the most formidable threat to the hatchlings and we made every effort to kill them. As on Europa, Casuarina and other small oceanic islands, access to water was a considerable problem for the rats and they lacked stamina. It was, as a result, relatively easy to chase them down and kill them. After three weeks or so we removed nearly 100 rats, but not without cost. One day I observed a large rat scrabbling at the base of a turtle nesting pit and I managed to slow it down with a well-aimed piece of coral. Confused, it ran past me and, being a sporting type, I took a mighty kick at it. My judgement was not perfect, unfortunately, and my big toe, regrettably unprotected by any footwear, encountered an outcrop of the base coral rock about 2 centimetres under the sand. I swear the entire 1,000-metre structure from the ocean bed to the surface resonated from the blow and I pitched howling into the nesting pit. Garth, not realising how near mortal was the damage, lost Brownie points by collapsing in a heap and laughing his head off. I was unable to do anything but writhe in pain. By the time he had sobered up and half carried me back to the Met station, my toe was twice its normal size and a more colourful and interesting globe I have yet to see. It was heavily streaked with green and blue bruises to which a veil of blood-stained sand adhered lightly. Our French companions were sympathetic, while having a good and justifiable laugh, and promptly produced from their medicine kit a new salve for such bruises. Laying it on thick, they bandaged my toe tightly and left me whimpering on the bed, where I stayed for two days unable to walk. On the third day, the toe was normal in colour and free of pain, an amazing testimony to modern medication. On returning to Durban I had the toe X-rayed and found that it had been broken clean across.

Garth, I am delighted to report, got a taste of my pain when he was chasing a ball and connected with a large coral rock himself. Although the damage to his toe was of a lesser magnitude than mine, the pain was satisfyingly sufficient, and I repaid him in kind with a good laugh. Happily, the damage to his toe was easily repaired.

During our visit, we covered the island thoroughly. We spent most of our nightly working hours on the west end, covering the kilometre of nesting beach, and our daylight hours on the reefs on the south of the island from where we would walk onto the bare coral flats occupied by a colony of masked boobies. These magnificent black-and-white birds nested in substantial numbers (we estimated that there were 300 nesting pairs) over most inhospitable bare ground. The full spectrum of nesting stages were there, from freshly laid eggs to newly hatched bundles of white fluff, to the spotty and grubby adolescent birds attempting their first flights. All the birds were remarkably tame and allowed one to sit next to them and even stroke them on the nest. Sadly, the nests were often associated with the skeletal remains of green-turtle females. If the females arrived too late in the morning to avoid being caught by the sun, they suffered heatstroke, became disorientated and wandered inland onto the booby nesting grounds, where they died of exposure. It does not appear to take much exposure to cause such disorientation.

The red-footed boobies, many of which nested in the *Tournefortia* bushes growing behind the nesting beaches, were similarly tame. If one approached very close, they would bob their heads in what appeared to be a threat, but no attack ever ensued. The body feathers of the red-footed boobies had either one or the other of two colour phases. One day, a nest of eggs would be incubated by a white-phase booby and the next day its mate, a grey-phase booby, would take a turn on the nest. The different colour phases appeared in some pairs and not in others. There did not appear to be any obvious pattern and the phases bred true without any shades between the white and the grey.

The boobies were, as on Europa, a primary source of food for the frigate birds, the other source being turtle hatchlings. A marked difference was noticeable in that the main food items regurgitated by the boobies were flying fish, whereas on Europa the main food item had been squid. The boobies did a good job of ferrying food for these restless predators as there were about 200 frigate birds on the island. They were not breeding when we were there and, unlike the boobies, the frigates were very wild and would not let you approach them at all.

The south coast was fringed with a healthy coral reef where we would snorkel on suitable days. Not only were the fish of interest, but on calm days it was possible to explore some of the deeper gullies without danger. Such forays occasionally proved invaluable as numerous artefacts of ancient shipwrecks were discovered. A large anchor and a number of severely rusted old cannons were embedded on the surface of one of the reefs. So, not unnaturally, it was tempting to contemplate the presence of some treasure lying hidden in the sand of the gullies. We did find treasure, but not of precious metal. My most exciting excavation unearthed a handful of lead grapeshot, which used to be fired from the cannons to rake the decks of the enemy. This handful of shot remains a precious possession.

In those days the meteorological staff used helium balloons to send up instruments for measuring weather conditions and we decided to adapt a few of these balloons to track the activity of green-turtle females after they had laid their eggs. This was a lot easier said than done. Green-turtle females, some exceeding 200 kilograms in weight, are capable of feats of amazing strength and speed. We used up quite a few balloons without getting a female into the water, with the originally enthusiastic meteorological staff beginning to panic at the loss of several balloons to the wild gyrations of the frustrated turtles. When we did eventually get a female and balloon successfully launched into the sea, the turtle swam along the north coast for precisely 650 metres before the lines parted company with the turtle, sending our last balloon on a futile flight into the blue sky.

We were thus unable to ascertain whether the green-turtle females spent their inter-nesting periods on the steep-sided flanks of Tromelin, the topography underwater there being remarkably more steep and rocky than the offshore sandbanks so common off the west coast of Europa Island. The nightly recoveries of tagged females certainly indicated that, once reaching the island, it is unlikely that they leave it until they have completed their nesting.

Males also converged on the island and we had splendid opportunities of witnessing and photographing mating behaviour as amorous males struggled for supremacy over a receptive female. They would pursue females with great vigour into very shallow water and would completely ignore us splashing around them to watch. Once in copula a mating pair would become completely oblivious of time and circumstance. On one occasion, we witnessed a pair left high and dry on the coral reef when the tide departed. They remained there, quite tolerant of our measuring and photographing, until several hours later the rising tide brought the water back to them and they clumsily swam off into deeper water. It was interesting to note that when stranded in shallow water, green turtles splash themselves with water using their fore-flippers. This happened far more often on Europa, where turtles were often caught by the falling tide on the sandbanks off the western beaches. On those occasions, the females would excavate a small pond with their fore-flippers, which filled with sufficient water to cover their heads and for splashing over their carapaces as it heated up during the day. Such activity appeared sufficient to prevent stranded females from overheating.

Tromelin was originally discovered in 1722 and was named Ile de Sable or Sand Island. It has no safe anchorage and was apparently avoided. However, in 1761 a slave vessel called *L'Utile*, carrying 80 black slaves, was wrecked on the island. The captain and crew built a crude flat-bottomed boat out of the wreckage and, leaving the slaves, made their way safely back to Madagascar. Faced with a constant shortage of water, the 80 slaves were left to live on what they could catch and kill, almost certainly including a healthy nesting population of green turtles. Fifteen years later, only seven survivors were found. All of them were women. In 1776 Captain Tromelin, of the ship *La Dauphine*, en route to Mauritius, spotted the castaways and, with great courage, brought his sailing vessel close enough to effect a rescue of the survivors. He brought them to safety in Mauritius and, in recognition of his courageous act, the island was renamed Tromelin.

This piece of history gave rise to some serious speculation as to why all the survivors were women, discussions which had reached no firm conclusions when we regrettably had to terminate our stay. I returned to Reunion and Garth to South Africa. My own research would take me on to Mauritius and the St Brandon islands, where women also played a dramatic role in the lives of the temporary inhabitants of this remote archipelago.

CHAPTER 28

# Among the Fairies: The St Brandon Islands

The final leg of the early expeditions was a visit to Mauritius in November 1971, which once boasted a thriving sea-turtle nesting colony, but is now, alas, only the recipient of the occasional turtle visitor. As with Reunion Island, sea turtles in Mauritius shared extinction profiles similar to the dodo and the solitaire, having played a valuable role in the survival of the earliest human inhabitants of the island. Fortunately, Mauritius is responsible for a number of isolated islands such as Agalega and St Brandon. Also known as the Cargados Carajos Shoals, but originally named São Brandão in 1546 (although no connection has ever been made with the Irish saint), St Brandon is situated some 450 kilometres north of Mauritius. It is essentially a 48-kilometre half-moon-shaped coral reef, bulging eastwards into the face of the South Equatorial Current. Scattered on and around this huge reef are some 26 small sandy islands, which are home to tens of thousands of sea birds and, once, probably similar numbers of sea turtles. The 1971 expedition was to try to ascertain what was left of the turtle population and to see what could be done to improve the conservation status of the islands.

The Mauritius fisheries authorities approved my visit and cooperated in every way. Mr Louis Couacoud, the delightful and enthusiastic owner of the Mauritius Fishing Development Company, made the St Brandon visit possible. He must have been aware that whatever information came back from the islands with me was likely to be presented in terms of nature conservation, a concept that, until recently, had received limited exposure in the islands and was almost certainly going to impact negatively on the way that the company had been doing business for many years. Despite his possible reservations, Mr Couacoud did everything possible to ensure that I managed to get a thorough look at the entire island group. Furthermore, he willingly made available company records of turtle catches and imports to Mauritius going back to long before the Second World War. Thanks to Mr Claud Michel of the Mauritius Institute and Museum, I was able to trace turtle records back into the nineteenth century. One of the admirable traits of British administrations is sound record keeping, which is invaluable to researchers.

After studying the data so generously made available, I was concerned that I might not find any turtles in the islands, but I was assured that there were still many turtles nesting there. Up to 500 turtles were being brought to Port Louis from the islands every year, which was a firm indication that they had not been exploited to extinction. I set sail on *La Perle II* with eager anticipation.

This soon changed to alarm as this particular ship was, I was informed by the skipper, a cross between a submarine and a corkscrew. It was narrow and rode very uncomfortably through the short, choppy waves of the Indian Ocean. By the time darkness had fallen, I was feeling distinctly ill and battling to maintain a grip on my dinner. By 8 p.m. I was desperate and the corkscrewing action eventually hurled me against the leeward gunwale, over which I damn near fell, saving myself from a watery grave only by grabbing a handy rope. My spirits, plus my dinner, left the vessel abruptly and I thought that life could not deal me a lower blow. I was wrong again.

The captain, Roland Tayolle, was a welcoming and friendly soul and he and the first mate observed my state with some amusement, but were sufficiently concerned to ask whether I wanted some anti-seasickness medication. At that point I would have agreed to anything, so I nodded and held out my hand for a pill (very British). To my surprise, I was turned around, my pants whipped down round my ankles and a suppository half the size of a rugby ball, lubricated with coco-butter, was thrust up my backside. This, I can assure you, will cure seasickness instantly. No one, in the long history of travel to St Brandon, has ever looked forward to arriving as much as I. The sight of the first islands the next day is etched into my consciousness.

Once we were in the shelter of the main reef, the sea was calm and we stopped to offload goods at several islands, eventually reaching Albatross Island, situated north of the main reef, just before darkness fell. Albatross could be seen from 20 kilometres away as it hosted an enormous population of nesting sooty terns, or *ye-ye*, a column of which rose above the island in a black cloud. The sheer volume of birds was astonishing and as I was landed near the base station, I could hardly hear myself think over the cries of the wheeling mass. The constant screaming made sleep virtually impossible and the situation was made slightly worse by the realisation that between *La Perle II* and the shore, half my supply of beers had disappeared.

There are no permanent residents on the islands and the Mauritius Fishing Development Company has the fishing rights for extended periods of time. To achieve their goals, a significant number of Creole fishermen (mostly Mauritian but some Seychellois), are contracted to work the concession from three of the larger islands: Albatross, Raphael and South. These 100 or more contract workers depart the islands each morning in smaller boats and catch fish using hand lines, which are brought back to the base islands where they are cleaned, salted and set out on racks or rocks to dry in the sun. Thereafter, the dried fish are bagged for transport back to Mauritius. The dried fish was more popular among the Mauritians than fresh fish because, from the salted fish, they could cut off what was required for a meal and the balance would still keep. No doubt, 40 years on, with greater prosperity, this situation may have changed, but there was no doubting the popularity of these crudely dried fish at that time.

The practice of contracting Creoles as fishermen has a history going back at least as far as 1850. The difficulties experienced with the labour force, even then, were recorded during formal visits by colonial administrators in response to complaints by fishermen of poor treatment. There were marked differences in opinion between the employer and the fishermen, naturally, and the early records show that the employers generally won. The fishermen's lives are hard and dangerous and I was surprised to find them cheerful and willing companions. Mr Couacoud arranged for me to have a sail boat with inboard engine and crew at my disposal, and for two weeks I travelled from Albatross Island all the way back to South Island, which turned out to be a most memorable experience.

As sea turtles obviously provided a welcome change in diet for the fishermen from time to time, I had some really well-informed guides. The Creoles were very knowledgeable about the green and hawksbill turtle, both of which nest in the islands. Loggerheads were well known, but mainly as juveniles that arrived seasonally, thus adding useful supporting evidence for our belief that they were arriving after having nearly completed a full circle in the south Indian Ocean gyre. Olive Ridley turtles and leatherback turtles were unknown, but they believed that there was another species, which they called either a *fahwa* or a *bakwa*. It was claimed that the animal was rarer than a green turtle and much larger and coloured bright green. I have absolutely no idea what it might have been, as the description does not match any known extant species.

The larger islands had concrete pens in which captured turtles were held until the next visit of *La Perle II*, but none were caught during my visit. As mentioned above, up to 500 green turtles made their way to Mauritius every year, over a very long period of time, and I am happy to say that this practice formally ceased in 1973. As for the Creole fishermen, however, I doubt that they would have terminated the practice of varying their diet with the odd turtle, but I am equally sure that the population could withstand such modest exploitation. During my two weeks on the islands, I counted at least 286 turtle nests, some old, but many recently made.

On the two occasions during our voyages that we saw turtles, one pair *in copula*, the harpoons were rapidly taken to hand and we gave pursuit. Happily, on one occasion, the engine broke down and that gave the turtle sufficient time to make itself scarce and, when chasing the mating pair, the female spotted us and dived into deep water. It was interesting to note that the local variant of the harpoon is called the *Johnkin*. Unlike the Malgache harpoon heads, the *Johnkin* head is simply a sharply pointed iron rod barely 10 centimetres long, without a barb. Once the harpoon is plunged into the carapace of the turtle, the shaft falls free and, by use of the attached cord, tension is steadily maintained and the harpoon head does not come out. When the turtle is manhandled into the boat, the harpoon head is quickly removed by pushing it hard against the carapace and then jerking it free. The resultant hole in the carapace is then plugged using a piece of wood or cork and when the fishermen return to their base island, the turtle, virtually unharmed, is placed in the turtle holding pens until the next arrival of the *La Perle II*.

We visited 20 of the islands, nearly all of which had tracks on the beaches, indicating that nesting was still taking place throughout the year. Although not one visit coincided with any nesting females, many tracks were clearly recent and I concluded that the wide scatter of the islands would ensure a degree of safety to many nesting females. Shortly after my visit, the Mauritian government, with the full cooperation of the Development Company, declared North Island a turtle sanctuary. If the response to protection observed in the French islands, and the loggerhead turtles nesting in South Africa, is anything to go by, then the easing of exploitation in the St Brandon islands will see the green-turtle population increasing steadily in years to come.

Every island that we visited, even those stripped flat by guano mining, was teeming with birds. Sooty terns were probably the most numerous species, but the delicate lesser noddy and slightly more robust common noddy were frequently noted on most islands. Common noddies were very aggressive and would not hesitate to attack if you approached their nests. They are fiercely possessive of their eggs and chicks; one female was observed incubating a large cowrie shell with intense determination! Frigate birds were nesting extensively on Grand Capitane island and red-footed boobies and masked boobies were frequently encountered. Many islands had large nesting colonies of roseate terns and when they rose glittering into the air they twinkled like jewellery against the azure sea. The sight was breathtakingly beautiful.

Among the numerous smaller species of birds, the most endearing was the fairy tern. Found on all the islands, it was a constant companion, flying around our heads as we sailed from island to island. Snow white, with boot-button eyes and blue legs, this delightful and almost ephemeral creature, similar in size to a laughing dove, is totally respected by all the fishermen because it is generally believed that if one is blown off course during a storm, the fairy tern will show you the way back to land. The fishermen had ample justification for placing their faith in the fairy terns as there were memorials to fishermen lost at sea on two islands and, only a few years before my visit, an entire island had been swept away by a cyclone.

The fairy tern has no fear of man and will happily permit you to photograph it to your heart's content without flying off. The respect of the fishermen is so deep that the fairy tern can get away with anything and will not be harmed. Being one of the most cavalier nesting birds on earth, there were many examples of the respect in which the tern is held. Fairy terns do not make a nest, but lay a single egg on whatever is handy. This can include bare branches, pieces of drying fish on the rack (which will not be touched until the chick has hatched and departed), drying washing placed on the ground and even, in one case, someone's hat. The latter nesting site was still occupied by the egg when we left South Island and the owner was happy to abandon the hat. The most vulnerable site I saw was the top of a mooring post driven into the beach. There was barely room for the egg and the sitting female, but in due course the egg hatched and the little tern held onto its limited world with a death grip. If it had fallen off during the day, it would have died within minutes on the searing hot beach. In the few days before leaving, I was delighted to see how the parent birds arrived with tiny fish, up to six at a time lined up in the beak, and carefully fed the chick while having only one or two toes touching the post.

While at Coco Island, I was invited by the fishermen to collect some rock lobsters. My colleague at ORI, Dr Paddy Berry, was an expert on crustaceans and I thought that it would be useful to collect a few specimens for him. At low spring tide we sailed to the main reef and then walked for somewhat more than a kilometre to the eastern edge where we awaited the incoming water. It was quite an eerie feeling standing literally on the edge of the world and looking out into the deep Indian Ocean knowing that the next piece of land was 7,000 kilometres away. In the twilight, the sea darkened to cobalt blue as the first slight waves crept over the edge of the reef. As darkness fell, the water started to sweep over our feet and with it came a flood of rock lobsters. No previous experience of rock lobsters prepared me for this phenomenon; they swept by in their thousands, occasionally flicking our ankles with their antennae. We did not even have to run around to fill sack after sack. This species, *Panulirus penicillatus*, is unique to island and offshore habitats and is only rarely found on mainland reefs. In Maputaland, it is occasionally encountered on Island Rock, a small outcrop of reef visible only at low tide about a kilometre offshore. The range of sizes was impressive, from small lobsters of barely 50 grams to huge specimens well over 2 kilograms. When one considers that our small group occupied barely 30 metres of reef and saw thousands of lobsters flashing by in the torchlight, the total numbers along the 48-kilometre reef boggles the imagination.

Back at ORI in Durban, Paddy and I, after some discussion, proposed a programme of controlled exploitation to the Mauritius Fishing Development Company in the belief that further detailed research should be undertaken and that if any regular exploitation was to take place, it should be in a planned, sustainable manner. Our offer of assistance was not taken up.

Coton, the chief fisherman guiding me around, took me to massive holes with gin-clear water in the middle of the reef, which were, to all intents and purposes, natural aquaria teeming with marine life. Some of the larger holes even had sharks swimming around in them. The water was so clear that on some occasions it appeared that the boat was simply floating on air and one could see shells and shrimps clearly on the sea floor. One day I asked how deep the water was, as I could see a beautiful cone shell below us. 'Several feet!' they cried and I stepped over the side, pipe in mouth, to retrieve the shell. I promptly went in well over my head, eventually having to turn and dive to the bottom several metres down. Fortunately, my pipe floated.

For men living in such harsh conditions, these fishermen made my stay a real pleasure, and in conversation one day I asked whether any women ever came to work the concession. They all said that this was not a good idea. Apparently, many years ago, the company decided to allow some fishermen to bring their wives to the islands. The endeavour had a tragic end as fighting soon broke out. The husbands refused to go fishing and leave their wives with non-fishing staff and eventually several fishermen died in knife fights and even some of the women were killed.

Any society devoid of women is sad, of course. The acutest memory I took away from St Brandon was of the manager of Albatross Island sitting with me and his illiterate cook who was hanging on every word read from a letter written by the cook's wife. The letter had arrived with me from Mauritius and it was a monthly task to read such letters to the recipients. The cook's look of rapt attention and sincere devotion was very moving. For those of us raised in a rather more sophisticated world, such a reaction to the spoken word is rarely seen and it was a special privilege to be a part of such an intimate family communication.

Accompanied, as the moment was, by the screams of thousands of sooty terns, it was a memorable experience indeed.

CHAPTER 29

# Beaches and Miracles

Beauty is said to reside in the eye of the beholder and when considering sea-turtle beaches I think that this applies to sea-turtle biologists. A common view among all those who study these creatures is that any beach is improved dramatically by the presence of nesting tracks. What has been a source of wonder to me, though, is the fact that sea turtles are pretty catholic in their choice and what may be, for many people, a most unattractive beach is often a site of choice for one species or another.

Sea turtles are tropical to subtropical in nesting distribution, so high-latitude beaches like those around Cape Town and the English Channel, apart from being miserable to swim in (even in the summer months), are decidedly uninteresting. The temperature is the important feature of successful nesting beaches. They should be warm enough to encourage incubation and the production of females, but cool enough, in parts, to ensure a constant supply of male hatchlings. Those criteria being met, turtles are remarkably tolerant of a wide range of beach colours, textures, aspects and widths. What is more, they are determined nesters and will surmount a range of obstacles to attain their goal.

In South Africa we have an abundance of beautiful beaches. The best are the almost endless, warm, golden silica-sand beaches from Cape Vidal northwards into Mozambique as far as the Paradise Islands. It comes as quite a shock, then, to visit beaches in Big Island, Hawaii, and the Atlantic coast of Costa Rica, to find that green turtles are content to lay in pitch-black volcanic sand. In contrast, many of the beaches around very tropical islands are pure-white coral sand and, at the head of the beach, this is as fine as flour in texture.

Even more surprising are the turtle-nesting beaches of the Guianas in South America. These beaches, especially in French Guiana and Surinam, migrate along the coast, driven by strong westerly currents that sweep up from the mouth of the Amazon River in Brazil. During my visit to Surinam, my colleague Dr Joop Schulz explained that it was necessary to build their base stations and hatching houses using temporary materials because every few years those built originally on the extreme western end of a beach find themselves at the far eastern end of a beach and were sure to be swept away shortly. The units would be dismantled and re-erected at the western end of the next available beach, which might be some kilometres away, separated from the departing beach by exposed mangrove roots without a grain of beach sand.

The power of the currents sweeping the Surinam coast is even more dramatically illustrated by the fact that thousands of hectares of coastal land and their associated beaches have been eroded away over the past few centuries. At low tide, the sea retreats for many kilometres, exposing extensive mudflats, which display a deeply incised, regular pattern of lines and squares. These are the marks of the deepest points of the canals originally dug in the seventeenth century to drain what were then the coastal wetlands. The richly productive coconut, sugar and other plantations that were made possible because of these drainage canals have long been scoured away, leaving only the scars on the mudflats to signal their passing.

The waters offshore are constantly dirty as the current carries millions of tons of silt spewed out of the huge rivers in the region. To me, the beaches were grubby and not picturesque, but thousands of olive Ridley, green and leatherback turtles come to the region every year. Quite how the nesting turtles adapt to the shifting beach regime may soon be understood, as a large team of French scientists recently started a long-term study of beach dynamics and sea-turtle nesting in French Guiana. The leatherbacks nesting in this region are as adventurous as those of Maputaland and it was from these colonies that the first transatlantic leatherback voyage was recorded; a female tagged in French Guiana was found off the coast of West Africa.

This exciting find was one of the first to demonstrate the migrating ability of leatherbacks and caused much excitement, but it was not half as exciting as the discovery of one of the largest leatherback colonies in the world, not far from the recovery site. Gabon was unknown territory for turtle biologists and it was only some years later that Jacques Fretey from Paris started working there and announced a nesting population of thousands of leatherbacks per season. At a time of growing pessimism over the long-term survival of leatherback turtles, this news was welcome indeed, especially after Gabon quickly proclaimed the nesting grounds a national park.

Ephemeral beaches are not restricted to the Guianas. When I was working in Cabinda, Angola, in 2005, near the mouth of the Congo just south of Gabon, I saw, courtesy of a Chevron helicopter flight, that the coastline was retreating and many of the beaches were narrow and in constant danger of being swept away. The olive Ridley turtles nesting there did not appear to have the privilege of stable nesting platforms and coastal erosion must have forced them to move north or south of their target beach during the nesting season.

Even in Maputaland we have seen modest examples of sand movement, none more dramatic than that experienced during Cyclone Claude in 1966 when a kilometre of beach south of Black Rock disappeared to a depth of 10–15 metres. John Bass and I discovered the missing beach when we fell over an embankment that had not been there the night before. A shift in the structure of the waves generated by the cyclone brought most of the beach back within three days. Happily, unlike the beaches of Cabinda and Surinam, this phenomenon did not disturb many nesting turtles because the affected area was quite limited.

In contrast were the tragic effects of the Indian Ocean tsunami in 2004. Kartik Shankar (who has been involved in turtle research in India for many years) described the disappearance of his entire leatherback-turtle beach, along with several of his staff, in the Andaman Islands, which were lifted nearly a metre by the earthquake. Kartik has recently reported that after six years the beaches are gradually returning, as are the nesting leatherbacks.

In my opinion, a tsunami must be old hat to a sea turtle. Turtles have survived millennia of natural disasters of this magnitude and, although they may be inconvenienced by the short-term disappearance of a beach, they will either find others or a reserve of younger animals will return to nest when a beach has recovered. The occasional turtle caught by surprise by the massive wave of a tsunami experiences plain bad luck. One only has to watch green turtles feeding on algae off a rocky headland during heavy storms to realise just how well adapted they are to living in the sea and how skilled they are in reacting to wave dynamics.

At the time of the Indian Ocean tsunami, I could not help being amused by a Sky News report on its side effects. Sky announced, couching this news in suitably maudlin and anguished tones, that the tsunami was threatening the sea turtles in the region with extinction. This ill-considered report appeared even less accurate due to the observations of a colleague from Sri Lanka. He told a gathering of us in Thailand that he had walked on a beach that had been swept over by a 10-metre wave during the tsunami two days before and had survived the dangerous withdrawal of the inundation. As he walked along the beach he was surprised and delighted to see the vigorous emergence of a clutch of turtle hatchlings, which successfully made their way into the sea. The popular press is all too prone to ignoring the evolutionary resilience of these amazing creatures and continues to depict them as vulnerable, delicate creatures requiring handling with kid gloves. Sea turtles, were they capable of listening to the utterances of many well-meaning NGOs, would be insulted by the references to their fragility. Sometimes hatchlings do not even require a beach.

One year, during a high spring tide that had been accompanied by a storm in Maputaland, we experienced a serious amount of beach erosion along the coast. This resulted in the exposure of many clutches of loggerhead eggs, many of which were lost as they rolled into the waves. The beach was littered with eggs for many kilometres. One clutch was found as the first egg tumbled out of the eroding sand bank and we hastily popped the egg into a plastic bag and added the rest of the clutch, which would have been doomed if we had left it. When we returned to Bhanga Nek, the plastic bag full of eggs was overlooked and one of the cleaning staff picked it up and put it in a safe place in the generator room. It remained there for some weeks until one night about midnight, as we were preparing for a beach-buggy patrol, we heard a scrabbling noise in the engine room. On investigation, the plastic bag was rediscovered, this time full of loggerhead hatchlings endeavouring to make their way out. The clutch had happily continued its incubation without a grain of sand around the eggs. One hopes, of course, that the warm generator room assisted in the emergence of a clutch of healthy females.

Many turtle beaches provide proof of miracles. A number of years ago, a publication stated that there was so much plastic waste on a beach near Hong Kong that nesting turtles could not scrape their way through it to dig a nest. I have no further details on that particular beach, but in 1988, when I spent some days on Ilha do Sal in the Cape Verde Islands in the Atlantic, I was utterly appalled at the litter on the nesting beaches there. It was possible to walk for hundreds of metres along a beach without setting foot on pure beach sand. These swathes of plastic and glass litter appear to emanate from Europe, much of it emerging from the Mediterranean Sea and being swept on to the Cape Verde Islands by the north-east trade winds. What is miraculous is that the nesting loggerheads do find a way through the litter and the hatchlings appear to emerge from the resulting nests and survive without damage from leached chemicals.

What is even more miraculous on Ilha do Sal is that loggerheads still arrive every year to lay their eggs. Along one of the main beaches, it was possible to walk over hundreds of square metres of turtle bones and carapace and plastron plates. These are all that remained of slaughtered turtles and had, over time, been flattened down into an almost piazza-sized testimony to the uncontrolled exploitation of this island population. The nesting loggerheads have been killed for food

*Beaches and Miracles*

on the island since it was settled in the fifteenth century and females were still being killed when I visited. If that is not a good example of the resilience of loggerhead populations, nothing is.

Happily, this resilience and patience is being rewarded. Conservation measures are now being implemented throughout Cape Verde and there are many dedicated people trying to ensure that this almost-forgotten loggerhead population will be restored again and be able to nest in peace in future. On Ilha do Sal this will not be an easy task, as there is little economic activity on the island. Its value as a source of salt declined sharply after the Second World War and the islanders have little access to other resources. One hope is that the Cape Verde islanders will respond in the same way as the people of Mohéli Island in the Comoros. The Mohélians are mainly fishermen, and also poor, and have little access to resources, but in the face of adversity they have adopted a conservation philosophy that is an example to the world.

Jack Frazier, having just completed an exciting period of study on Aldabra (having been seduced away from his primary doctoral study of Aldabran tortoises by the nesting sea turtles), took a break in 1972 and sailed solo in a small yacht from the Seychelles to South Africa, calling in at numerous islands en route, one of which was Mohéli. He described the existence of a hitherto unknown, and large, nesting colony of green turtles on the beaches of Itsamia on the south-east coast of the island.[1]

In 1987 the South African Department of Foreign Affairs supported a visit of specialists to the Comoros. Alongside Dr Mike Bruton, director of the JLB Smith Institute of Icthyology in Grahamstown, and Dr Colin Buxton, a marine-fish authority, I had the good fortune to spend two weeks in the archipelago, visiting the three main islands of Grande Comore, Anjouan and Mohéli. It was a wonderful visit, especially to Mohéli and the village of Itsamia, where we found the village community extremely proud of their nesting green turtles. They were determined to protect them, not an easy task because there were many poachers coming over from neighbouring Anjouan where nesting turtles were now exceedingly rare. We left Mohéli convinced that we should provide Itsamia with as much support as possible and in our report we recommended to the Comoros government that they declare the beaches in the south of Mohéli a national park. It took some years and the assistance of many volunteers and NGO bodies, but in 2001 this beautiful area was declared a national park, the first in the Comoros. The Itsamians have received many accolades for their conservation endeavours and their beaches are attracting tourists from overseas and scientific help and support from Ifremer and Kelonia in Reunion. In 2012 the government of the Comoros established a formal and permanent police presence adjacent to Itsamia to add support to the wonderful efforts of the local villagers.

In a small, poor country with minimal resources, what has been achieved on this south-east corner of Mohéli is nothing short of a miracle as over 3,000 green-turtle females a year are nesting on this very special beach. Contrast their actions with Grande Comore, where nesting turtles have ceased to exist; Anjouan, where they are now a rarity as a result of widespread killing; and the beaches in the north of Mohéli, where the killing of nesting turtles is still commonplace.

One local Mohélean taught the expeditionary group a modest but effective lesson in humility. Mike, Colin and I all held doctorates, a combined higher-education commitment of over 21 years. If one adds our combined schooling, a total of 57 years of education had brought the three of us to where we were. While driving to Itsamia in a hired and extremely dilapidated Renault 4L, we found our way blocked by a fallen tree. The road was narrow with steep banks and there was no way of driving around it. We were determined not to have to drive back around the island to get to our destination. So, having neither axe nor machete, we leapt out and with a combined effort tried to drag the tree off the road. We went at it, bull at a gate, and just about ruptured ourselves without success. After ten minutes of really serious endeavour we were exhausted and were sitting, rather disconsolate, contemplating our next move.

Suddenly, there appeared a single villager who took in the situation at a glance and without a word turned and disappeared into the bush. Moments later we heard some chopping and within minutes the villager reappeared carrying a freshly cut pole. Without a sound he walked up to the tree and, with a few simple and effortless moves, and minimal help from us, simply levered the tree off the road.

Wordlessly, he tossed the pole away and strolled off without waiting for any thanks or reward and left three highly educated and totally humiliated biologists sitting stunned next to the car.

It was, therefore, a small miracle that we made it to the beach at all.

1 Frazier, J. 1985. *Marine Turtles in the Comoro Archipelago*. Amsterdam: North Holland Publishing Company.

CHAPTER 30

# The Sodwana Declaration

In 1995 the Natal Parks Board hosted a sea-turtle workshop under the auspices of the IUCN and funded by the Convention of Migratory Species (CMS), the United Nations Environmental Programme (UNEP) Water Branch and the WWF. This initiative was a result of the sterling endeavours of Rod Salm from the IUCN office in Nairobi and Douglas Hykle, an office-bearer of the CMS who developed a passion for sea turtles. The choice of Sodwana Bay was not accidental as, at that stage, our turtle programme had been going for over 30 years and had become well known in the international world of sea-turtle conservation.

Nothing is more rewarding than being recognised by one's peers and there were two memorable incidents which made all the effort worthwhile at that time. One was a newspaper article in the *Malaysian Times* in which two turtle researchers claimed that they were determined to keep working to save what appeared to be their fast disappearing population of leatherbacks in Terengganu because they were inspired by the work being carried out in South Africa. This absolutely delightful recognition occurred before the Sodwana gathering. The second was an announcement by Col Limpus, at a conference after the workshop, that the only programme he was confident had demonstrated a positive response to intense conservation endeavour was 'on the beaches of Tongaland (*sic*)'. Such recognition was, and still is, worth its weight in gold, even though, 20 years later, many other colonies of turtles have growing populations.

It is a matter of great pride that our loggerhead nesting population has multiplied four times over the past 50 years and is likely to expand further as the southern Mozambique beaches are now enjoying the same high level of protection. The leatherback population has not responded as well and there may be many reasons for this. Apart from being the most southerly nesting colony in the world, with many of the hatchlings being swept down into the waters around the Cape, leatherbacks face the pelagic sea-fishing techniques such as drift nets and long lines, which have impacted seriously on leatherback populations.

Conservation, be it aimed at sea turtles or any other species, is a long-term affair with broad horizons, and it needs participants who are excited by the challenge, identify with the animals and are willing to put in a massive effort in order to achieve their goals. Sea turtles appear to have the inherent ability to create the ambiance within which all these admirable qualities thrive. Since the beginning of the sea-turtle protection programme in Maputaland, the turtles have woven their magic spell on scientists, students and field officers alike.

The first ten years of my time in Maputaland were shared with some of the most enthusiastic and dedicated people one could ever meet. The game rangers who took their turn on the beaches as officers-in-charge are, to a very large degree, the reason why such good and accurate results were obtained over the years. Not only did they start the annual programmes before the scientists and students arrived and complete the programmes after they left, but it was their task to prepare the station, employ the temporary staff, organise the permanent staff, repair the cottages (or build them), lay on the water and get the beach vehicles into a condition that would endure a trying season. I have already mentioned some of the additional duties that fell to their lot, such as looking after horses, hand grenades, military personnel and local community leaders.

The most time-consuming task was the vehicle patrols that covered over 100 kilometres of beach every night. Although this sounds a simple matter, in Maputaland it is not. The beaches are dynamic, have varying sand textures and steepness and can be unkind to the unwary or tired driver. There is a constant threat of the vehicle becoming bogged down and promptly being in danger of being washed away by an incoming tide. Losing vehicles on the beach is a career-limiting achievement. All of these pressures can take a toll on the local officers and their state of mind was often neither appreciated nor recognised.

Of course, sometimes these pressures are self-inflicted. At the beginning of one season, Herman Bentley collected a brand new beach buggy and decided he would drive it up to Bhanga Nek along the beach from Sodwana Bay. Once on the beach, he encountered heavy rain, which pelted him directly, as beach buggies in those days had no roof. So, he decided to speed up to get home. The combination of heavy rain and lack of direct vision resulted in Herman missing a turn at a rocky headland and leaving the safety of the sand. He became aware that all was not well as the sound of the beach changed from the steady and comforting hiss of sand under the wheels to a rattling rumble as he took the first bed of acorn barnacles on the reef. Applying the brakes too late, he scored a pair of tracks (that remained visible for years afterwards) across a bed of acorn barnacles and plunged into a rock pool.

Herman, an irrepressible man of unquestioned commitment, ran for help, returned and had the buggy towed to Bhanga Nek, where he personally stripped the engine, cleaned every part, reassembled it and had the buggy in tip-top condition for the beginning of the year's survey. Now that is dedication.

The game guards, now named field rangers, both permanent and temporary, of Bhanga Nek are in a league of their own. It was Mbika, Sonto, July, Onions and many, many more, over the past 50 years, who kept the turtle programme going. In the early days it was the guards who maintained the vehicles and the boats and accompanied the officer-in-charge everywhere, looking after the station when he was absent. They maintained discipline among the temporary staff, provided guidance and help for the students and ensured that the daily data notebooks and logistic necessities were provided to the field camps in need. Without their willing, cheerful and committed endeavours, the turtle-conservation programme would never have been a success.

Finally, of course, the achievements of the programme would have been considerably less dependable had it not been for the wives of the officers-in-charge. These ladies often took responsibility for ensuring that all the nesting and morphological data gathered every night was transcribed into the record books and checked and checked again. I should mention that we all used pencils and paper for many years and it was only with the arrival of senior ranger Noël Wright that the computer made its debut on the beaches. Nowadays this tool is ubiquitous, but for many years it was good old-fashioned writing that maintained the records. What an incredible debt we owe these women who often sat with us late into the night marking hatchlings or gathering eggs and whose enthusiasm matched and, on occasion, exceeded that of their exhausted husbands. It should be said that many a student, myself included, remains eternally grateful to the wives for their more than occasional decent meal and their medical care and tender commiseration during times of accident.

The Sodwana Bay workshop was, therefore, an important milestone. It was a product of hard work by a kaleidoscope of people from rural labouring staff, highly experienced field officers and game rangers and their families, to professional scientists and carefree students. If any one of these categories was taken out of the loop, the results would have suffered.

The Sodwana meeting was so important to staff of the Natal Parks Board that posters, beautiful T-shirts and models of turtle hatchlings were produced by the Parks Board's design studio. The hospitality and catering were unbelievable. The entire effort made a huge impression on our visitors, not the least of whom was the minister of Traditional and Environment Affairs, Inkosi Nyanga Ngubane, whose appointment dated from 1994.

South Africa had, only the year before, undergone a dramatic political transition and, as our first Zulu provincial minister associated with the environment, Inkosi Ngubane could not have had a more exciting opportunity to acquaint himself with the sea-turtle programme. He also obtained a first-hand understanding of what had been achieved in this field and how much the effort was appreciated by the international conservation community. He loved every moment of his participation and became one of the finest supporters of conservation the province has ever had. The Sodwana gathering exposed Inkosi Ngubane to practical research and conservation at grass-roots level and he glowed with pride in his conservation organisation. This pride remained with him throughout the five years that he held his political portfolio and was obvious until he died in 2010.

My own circumstances changed about the same time. Retirement beckoned and the long-held direct association with the turtle programme was coming to an end. One of my more pressing concerns was that of succession. Would the turtle programme find a new caretaker? As CEO I had (perhaps with a modest display of bias) managed to maintain a steady flow of financial support for the programme. I had also occasionally guided staff to ensure that if there was a shortage of funds (which there often was) the turtle programme should not be the one to suffer terminally. May I add that I feel no shame for my behaviour as everybody has a weakness and, you may have noticed, sea turtles are mine.

The gods, at this time, noted that I was in need. This time, obviously recognising the need for gender equality, they sent a woman. Ezemvelo KZN Wildlife appointed Dr Ronel Nel as a marine scientist and part of her extensive portfolio was overseeing the sea-turtle programme. With a PhD in beach dynamics, and thus a direct connection with sea turtles, she took a deeper than expected interest in the Maputaland programme. In less than a year, Ronel developed a passion for the programme and the sea turtles and rapidly applied her modern skills and talents to improving the beach procedures and recording techniques. She has become a joy to me and has not only rejuvenated the research but has, over the last decade, become one of the foremost turtle-research scientists in the world. With the help of her students (not only those she employed when she worked for Ezemvelo KZN Wildlife, but her postgraduate students since her return as a full-time lecturer at the Nelson Mandela Metropolitan University) she has become a world authority in the field.

## The Sodwana Declaration

At a recent conference in Australia, I admitted that the apparent good fortune of encountering someone to whom I could sedately pass the baton did not quite fit the facts. I was bumbling around holding the baton until Ronel arrived and thankfully tore it out of my hands. She and her colleagues have done an outstanding job. They have not only ensured that the programme sails on with added impetus, but have taken the pencil records and notes of nearly five decades of work and organised and analysed them to produce some of the most truly definitive analyses of sea-turtle biology ever published. She has made South Africa proud as she has intensified and expanded the research programme and has made me proud by honouring me with her friendship.

At the end of the 1995 Sodwana workshop, all the attendees present agreed on the wording of a declaration summarising what they collectively felt was necessary to achieve success in sea-turtle conservation in the western Indian Ocean. The document is called 'The Sodwana Declaration'. Among the conclusions reached are the following:

- 'The marine turtles of the Western Indian Ocean are a shared resource of inestimable value to the region's coastal nations.'
- 'The long-term survival of the marine turtles of this region cannot be achieved by any single country and maximum collaboration and cooperation are necessary.'
- 'The popularity of marine-turtle nesting areas as exciting ecotourism destinations makes them a valuable economic resource, the use of which, through rational non-extractive methods, is both laudable and sustainable.'

Nearly five decades of research have proven the truth of the first of these conclusions; not only are our South African loggerhead and leatherback nesting populations shared between the western Indian Ocean countries, but also with those along the eastern Atlantic Ocean. The green turtles of the western Indian Ocean, from a myriad of well-protected islands, intermingle freely over the vast feeding grounds associated with the Mozambique Channel. It is as a result of having the nesting grounds so well protected that it is still possible for the Vezo, Sakalava and other Malgache coastal tribes to harvest their beloved green turtles. Every reef in the entire region hosts small numbers of hawksbill turtles that continue to survive against all odds. What is awaited with interest is confirmation of where the modest populations of olive Ridley turtles emanate from. They may even be Indian in origin, as no nesting ground has yet been found in the south-western Indian Ocean. The possibility that they nest further afield is not far-fetched as sea turtles are astounding wanderers. In recent years Kelonia has satellite-tagged numerous loggerhead sub-adults caught offshore and demonstrated that many are returning to Omanese waters and probably originate from the great loggerhead nesting colony on Masirah Island, Oman, many thousands of kilometres to the north.

The second conclusion in the declaration is self-evident. It is a matter of some pride to all South Africans associated with the Maputaland programme that, not only because of the Sodwana workshop, we have lent a hand wherever possible to other countries in the region and even as far afield as Bangladesh. We would like to think that our endeavours have contributed to the declaration of no less than ten protected areas centred on sea-turtle nesting colonies, many of which are recording growing nesting numbers. As a result, the future survival of many thousands of sea turtles is a great deal more assured than it was five decades ago. It is to be hoped that there are more protected areas to come.

The relatively recent establishment, in 2001, of the Memorandum of Understanding for the Conservation of Sea Turtles in the Indian Ocean and South East Asia (IOSEA) by the CMS, led by Douglas Hykle, has expanded the vision of the Sodwana Declaration to the entire Indian Ocean region. South Africa is a valued signatory member.

The third conclusion, endorsing the benefits of ecotourism associated with nesting turtle colonies, has proved invaluable, and not just in South Africa. In Mozambique, Madagascar and especially on the beaches near Itsamia in Mohéli, tourism is a very positive force and more of it will be welcomed by everyone. Tourists love sea turtles, which is not surprising. These fascinating animals have wandered the seas of the world for millions of years and the sorry state to which many populations have been brought, is a direct result of the stupidity and cupidity of humans. It is, therefore, only right that humans should try to restore the damage done to an undeserving victim over the past 500 years.

What an honour and pleasure it has been to share this field of endeavour with so many thoroughly decent and talented people, starting with the legendary Archie Carr, in conference halls and turtle beaches all around the globe. The efforts of each of them have been constantly aimed at assuring the survival of sea turtles.

The first wave of turtle biologists and conservationists is passing and I wish all strength to the younger enthusiasts carrying on with research and protection. I applaud their efforts and wish them well, for no reason, perhaps, other than the fact that I owe sea turtles an enormous debt for what they have given me: joy, excitement, fulfilment and reward. In addition they have generated in me a near lifetime of deep affection.

I hope that I have done them some good.

# Photo Captions

1. Leatherback turtle.
2. Olive Ridley turtle.
3. Green turtle.
4. Hawksbill turtle.
5. Loggerhead turtle.
6. A truckload of schooner-caught green turtles in Grand Cayman, 1974.
7. A loggerhead female in her body pit in Maputaland, 2009.
8. Sea turtles scoop out sand from their nest chamber using their hind-flippers like human hands.
9. A soup label from the 1960s. Photo by courtesy of Bon Vivant Soups.
10. Humph McAllister, the first student at Bhanga Nek, with a loggerhead-turtle skull, 1964. Photo by courtesy of O. Bourquin.
11. A koi made from tortoiseshell – a fine example of the art of *Bekko*. Tokyo, Japan, 2011.
12. A scene we came upon in 1963, which caused this part of the Maputaland beach to be named 'The Graveyard'. Photo by courtesy of the Natal Parks Board.
13. Garnet Jackson, who rebuilt the Bhanga Nek cottage after the helicopter disaster, 1985.

*Photo Captions*

14. The author with one of his first loggerheads, near Bhanga Nek, 1965. Photo by courtesy of the *Natal Mercury*.
15. The first meeting of the IUCN Marine Turtle Specialist Group in Morges, Switzerland, 1969. Bob Bustard, the author and Peter Pritchard are in the back row, with the chairman, Archie Carr, second from the right in the second row.
16. Brian Stevens, the author and Onions Gumede during the author's first season on the Turtle Survey in 1965. Photo by courtesy of the *Natal Mercury*.
17. Mr Peter Potter – the father of the Maputaland Turtle Survey, 1985. Photo by courtesy of the Natal Parks Board.
18. The wreck of the *Rocktail*, south of Black Rock, 1966.
19. Major Pat Temple and the famous horse guards at Bhanga Nek, 1966.
20. The Malangeni River in flood, 1966. Mike is in the river, John and Mbika are on the Land Rover.
21. The normal tagging site on loggerheads – on the trailing edge of the fore-flipper as near to the body as possible – showing a titanium tag in place.
22. The various tags used on turtles in Maputaland from 1963–80.
23. Monel tags removed from re-migrants, some of which show extreme corrosion.
24. Drs Paolo Luschi and Alessandro Sale make final adjustments to a satellite transponder fitted onto a leatherback, south of Bhanga Nek, 2003.
25. The first attempt at marking hatchlings at Bhanga Nek using trout tags, 1967.
26. Tagging loggerhead hatchlings at Bhanga Nek with stainless steel wire, 1971.
27. A basket of loggerhead hatchlings at Bhanga Nek waiting to be tagged, 1980.
28. A leatherback hatchling. Note the scales, which disappear completely as the animal grows from 60 grams to nearly 1,000 kilograms.

29. The first notching of a loggerhead hatchling in Maputaland, 1972.
30. During the post-hatchling stage, loggerheads develop sharp protective spines on their vertebral scutes, which are lost as the animals mature.
31. The Panda Wagon meets its first loggerhead at Bhanga Nek, 1969.
32. Senior Ranger Barry Brent, who erected the first distance beacons along the coast, crossing Lake Hlange, 1969.
33. Professor Archie Carr, a pioneer of turtle conservation, at the Tortuguero turtle base camp in Costa Rica, 1983.
34. Jeff Gaisford, Retired Media Officer of Ezemvelo KZN Wildlife, one of the best friends that that turtles ever had. Photo courtesy of Catherine Hughes.
35. The mouth of the Kosi Bay Estuary, which shows the density of the fish traps, 1966.
36. A clutch of loggerhead eggs in Maputaland, 2003.
37. Ranched green turtles in the Mariculture Ltd turtle farm on Grand Cayman, 1979.
38. David Rowe-Rowe and the author sampling loggerhead eggs in Maputaland, 1967.
39. Transferring eggs en route to Bhangazi North beach, 1985.
40. The dry coral sand on Casuarina Island off the coast of Mozambique necessitates the digging of deep body pits, 1970.
41. The Panda Wagon driving the beaches near Nacala Porto, Mozambique, 1970.
42. Yves Riou, a local artisan skilled in working tortoiseshell, demonstrates his trade at Kelonia, 2010.
43. The impressive entrance to the Kelonia Institute in Reunion, 2012.
44. The main tank at Kelonia on Reunion Island, 2007.
45. A turtle release accompanied by singing children at Kelonia on Reunion Island, 2003.

*Photo Captions*

46. Altars, which reveal the respect accorded to sea turtles, were once common in Vezo coastal villages in Madagascar, 1970.
47. Phillipe and Renée, two of the meteorological staff, weighing a 210-kilogram green-turtle female on Europa, 1970.
48. Tromelin Island appeared to be all airstrip, 1971.
49. Jose Tello at Inhassoro, Mozambique, 1969. These are the remains of turtles caught in the seine nets of Inhassoro.
50. Paul Dutton with the 'Spirit of the Wilderness' after a bumpy landing on the beach to check a leatherback nest, 1968.
51. A masked booby chick amidst turtle bones on Tromelin Island, 1971.
52. Courting green turtles off Europa, 2010. Photo by courtesy of Jerome Bourjea.
53. Green-turtle nesting tracks during summer on Europa, 1971.
54. Courting green turtles at dawn off Europa, 2010.
55. The remains of a poached green turtle on Moheli, 1987.
56. A typical fishing boat leaving Raphael Island, 1971.
57. The *Antsiva* lying off Juan da Nova, 2010.
58. Interviewing villagers always attracted a great deal of interest, Bimbini, Anjuan, Comores. 1987. Photo by courtesy of Colin Baxton.
59. Senior Ranger Garnet Jackson with his student team at Bhanga Nek, 1985.

# Acknowledgements

It is not possible to thank every individual or organisation that has made it possible for me to write up these memoirs of an enjoyable and rewarding association with the sea-turtle family. If I tried, half the book would be names. It should be acknowledged, however, that I owe everyone who helped me on my way an enormous debt of gratitude and I thank them profusely, one and all, for their support, friendship and encouragement.

Where possible, and with great pleasure, I have included in the text the names of many who guided and helped me along this wonderful voyage. There are many more but space precluded their inclusion. Some of those deserve special mention.

My parents, Mitchell and Connie Hughes, both true-born Scots from conservative sea-fishing families, brought their family to South Africa and soon feared for the future of their second son. They became seriously bewildered when he plunged into nature conservation. When snakes entered the house in the inexperienced hands of this amateur naturalist, mild terror accompanied the confusion. My noble young sister Marybelle gave them comfort and reassurance, for which I am eternally grateful.

## Acknowledgements

My many friends and colleagues in the Natal Parks Board, especially Bill and Leila Barnes during my four years at Giants Castle, gave me encouragement, support and the will to take the plunge into higher education. All the Natal Parks Board staff on the turtle beaches, including officers, local guards and students, what a wonderful contribution you have made and what pleasure you have given me over the years. Colonel Jack Vincent and John Geddes-Page, the first two directors of the Natal Parks Board, must be thanked for repeatedly finding me employment.

My scientific and technical colleagues at the Oceanographic Research Institute gave me five years of sheer unmitigated joy. They exposed me to hard work, scientific debate and hysterical situations and I wish I could have told more of the incredible stories from that time.

Then, in the dark hours of unemployment, when I was facing an uncertain future and near poverty, Jenny and Brian Hutchinson and then Mike Poulter found me a bed until Pat Acutt opened his old Wendy house.

Dr Allan Heydorn has my thanks for agreeing to write the preface and I am encouraged to see that his spirit of optimism and forgiveness is still apparent. He always was an innocent.

May I gratefully acknowledge the permission given by Mrs Julia Coetzee to use one of Mark Coetzee's creative works; the Department of Environmental Affairs and Dr Paolo Luschi, University of Pisa, for the use of their satellite images; Piet van As for allowing the inclusion of a few of his memorable and appropriate cartoons; and finally Diana Martin for her recreation of the maps.

To Carol Broomhall of Jacana Media, my deepest gratitude for providing the opportunity to start writing in earnest, and to Pete van der Woude for his patience in editing and his many useful suggestions and corrections.

Finally, thanks to my family, Lee, Catherine and Mitchell, who suffered my long absences with fortitude and became involved in the whole turtle saga themselves. Their enthusiasm and encouragement brought me both joy and justification. In recent months, Lee's thorough and encouraging editing of my first attempts at this book spurred me on. Without their love and support, my life would have been neither so interesting nor so rewarding.

It is often said that there is no such thing as luck. Well, that is simply untrue. So many good people have created and nurtured mine and I hope that they share the pleasure that they have helped to generate.

*George Hughes*
*Howick July 2012*